A REVISED KEY TO THE
BRITISH SPECIES OF

CRUSTACEA: MALACOSTRACA
OCCURRING IN FRESH WATER

with notes on their
ecology and distribution

by

T. GLEDHILL, M.I.Biol.,
D. W. SUTCLIFFE, Ph.D.,
Freshwater Biological Association

and

W. D. WILLIAMS, D.Sc.,
Department of Zoology, University of Adelaide, Australia.

FRESHWATER BIOLOGICAL ASSOCIATION
SCIENTIFIC PUBLICATION No. 32
1976

PREFACE

As in the earlier work published by the Association as *Scientific Publication* No. 19, this key is basically concerned with species of Malacostraca known to breed and live permanently in fresh water in Britain. Also included are a few species which breed in brackish water but nevertheless are regularly found in fresh water, at least in certain localized areas. In addition the key to the genus *Gammarus* embraces all species present in Britain, including the marine-littoral species, since there is no modern published key to the British members of this genus and nearly all may be encountered by the freshwater biologist working in coastal habitats. Mr T. Gledhill and Dr D. W. Sutcliffe compiled this new publication, Mr Gledhill having special responsibility for *Bathynella* and *Niphargus*, and Dr Sutcliffe for crayfishes and *Gammarus*. Professor W. D. Williams revised his key to *Asellus*. All of the keys include one or more additional species, but the present work owes much to the groundwork laid in the earlier publication by Professor H. B. N. Hynes, and by Dr T. T. Macan who was responsible for initiating both editions.

Distribution maps have not been included because the distribution of both amphipods and isopods is under review elsewhere. The British Isopoda Study Group operates a recording scheme (Isopod Survey Scheme) and records can be forwarded to Dr P. T. Harding, Monks Wood Experimental Station, Huntingdon. In collaboration with the Biological Records Centre of the Institute of Terrestrial Ecology, the B.I.S.G. intends shortly to publish maps for the genus *Asellus*. A survey of amphipods has also been started, again using cards supplied by the Biological Records Centre. In place of distribution maps a more detailed bibliography has been included, to assist those readers who may wish to find details of recent work on the distribution of freshwater malacostracans. It is hoped that the bibliography will also serve as an introduction to the extensive literature dealing with other aspects of the ecology of this group.

SBN 900386 24 X

CONTENTS

INTRODUCTION	4
THE KEYS	7
Syncarida: Bathynellacea	7
Eucarida: Decapoda	12
Peracarida:	22
Mysidacea	23
Amphipoda	24
Isopoda	58
ACKNOWLEDGMENTS	65
REFERENCES	66
INDEX	72

INTRODUCTION

The Sub-class Malacostraca are Crustacea which have retained eight thoracic and six (seven in the marine Leptostraca) abdominal segments. In addition to the abdominal segments there is a tailpiece or telson. The embryonic head contains six segments, but the adult head contains only five segments. Typically a carapace encloses the thorax. The female and male reproductive openings are on the sixth and eighth thoracic segments respectively. Figure 1 illustrates a generalized malacostracan. Most of the trunk segments bear appendages which exhibit considerable variety of form, but in the freshwater species they are built up on the two-branched pattern and in no case form long series of similar foliaceous limbs such as are characteristic of the Branchiopoda. The terminology applied to the thoracic and abdominal appendages of malacostracans is confusing. The first one to three pairs on the thorax may be called *maxillipeds*, the second and third pairs are in some groups termed either *walking legs* or *gnathopods*, and the last four or five pairs are often called *walking legs*. Figure 2 (see also fig. 6C and D) illustrates a generalized thoracic appendage or *pereiopod* (=peraeopod). On the abdomen the first three pairs of appendages are usually referred to as *pleopods* and the sixth pair are termed *uropods*, as are sometimes those on abdominal segments four and five. The terminology used here in the keys for each group follows that shown in Table 1.

The text-figures usually illustrate the left-hand side of the whole animal, and appendages or parts of the animal are also usually drawn as seen when viewed with the head of the animal to the observer's left.

The Malacostraca are divided into five super-orders: Phyllocarida, Syncarida, Hoplocarida, Peracarida and Eucarida. All members of the Phyllocarida and Hoplocarida are marine. The super-order Pancarida, which contained the order Thermosbaenacea, has been omitted as the latter is now generally recognized as a separate order of the Peracarida. Recently, Serban (1970, 1972) removed the order Bathynellacea from the Syncarida and placed it in a new super-order, the Podophallocarida. The Syncarida and Eucarida contain only a few freshwater species, and these are easily recognized from illustrations. They are therefore described first without the use of formal keys down to genera.

INTRODUCTION 5

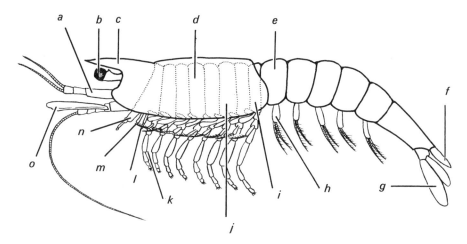

Fig. 1. Lateral view of a generalized malacostracan (after Calman). *a*, antenna 1; *b*, eye; *c*, rostrum; *d*, carapace; *e*, abdominal segment 1; *f*, telson; *g*, uropod; *h*, pleopod; *i*, male segment; *j*, female segment; *k*, pereiopod 1 (first thoracic appendage); *l*, maxilla 2; *m*, maxilla 1; *n*, mandible; *o*, exopodite or scale of antenna 2.

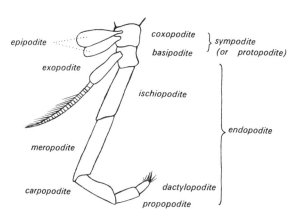

Fig. 2. A generalized thoracic appendage or pereiopod (after Calman).

Table 1. The names of appendages in some malacostracan crustaceans

	Bathynella	Astacura	Gammaridae	*Asellus*
HEAD	Antenna 1 Antenna 2 Mandible Maxilla 1 Maxilla 2	Antenna 1 Antenna 2 Mandible Maxilla 1 Maxilla 2	Antenna 1 Antenna 2 Mandible Maxilla 1 Maxilla 2	Antenna 1 Antenna 2 Mandible Maxilla 1 Maxilla 2
THORAX	Pereiopod 1 Pereiopod 2 Pereiopod 3 Pereiopod 4 Pereiopod 5 Pereiopod 6 Pereiopod 7 Pereiopod 8	Maxilliped 1 Maxilliped 2 Maxilliped 3 Cheliped Walking leg 1 Walking leg 2 Walking leg 3 Walking leg 4	Maxilliped Gnathopod 1 Gnathopod 2 Walking leg 1 Walking leg 2 Walking leg 3 Walking leg 4 Walking leg 5	Maxilliped Walking leg 1 Walking leg 2 Walking leg 3 Walking leg 4 Walking leg 5 Walking leg 6 Walking leg 7
ABDOMEN	Pleopod 1 Pleopod 2 Pleopod 3 Pleopod 4 Pleopod 5 Uropod	Pleopod 1 (gonopod ♂) Pleopod 2 (gonopod ♂) Pleopod 3 (swimmeret) Pleopod 4 (swimmeret) Pleopod 5 (swimmeret) Uropod	Pleopod 1 Pleopod 2 Pleopod 3 Uropod 1 Uropod 2 Uropod 3	Pleopod 1 (absent in ♀) Pleopod 2 Pleopod 3 Pleopod 4 Pleopod 5 Uropod
TELSON	Telson	Telson	Telson	Telson

THE KEYS

Super-order SYNCARIDA Packard, 1886
Order BATHYNELLACEA Chappuis, 1915

BATHYNELLA is a genus of small, c. 1 mm long, eyeless, more or less colourless subterranean crustaceans. The body is elongate, the carapace is absent and the thoracic limbs except the last have two branches. All but the first and last abdominal segments are without appendages. The first antenna is unbranched, the second has a small branch. Figure 3 illustrates the form of a typical bathynellid.

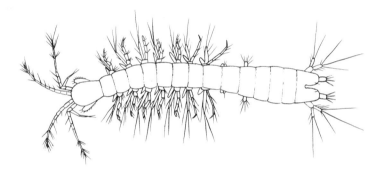

Fig. 3. *Bathynella natans* (after Thienemann).

Until Serban & Gledhill (1965) recorded *Bathynella natans stammeri* Jakobi from England only one species, *Bathynella natans* Vejdovsky, had been recorded (see Gledhill & Driver, 1964). Serban (1966b) separated *Bathynella stammeri* Jakobi from the *Bathynella natans* Vejd. sensu Jakobi group and placed the former in a new subgenus *Antrobathynella*. Similarly, Husmann (1968) showed *B. natans* Vejd. and *B. stammeri* Jakobi to be separate species. Serban (1973) raised the subgenus *Antrobathynella* Serban, 1966 to full generic rank and suggested that the different populations of *stammeri* do not belong to a single species.

1   Mandible with 7 teeth (pars incisiva 3, pars molaris 4), (fig. 4A–C);
    uropodal sympodite with 5–7 (frequently 6) spines (fig. 4G)—
    Bathynella natans Vejdovsky

Pereiopod 8 of male with anterior plate triangular, well developed and without spines on internal surface; internal lobe cylindrical, not extending beyond distal limit of anterior plate; small lobe vermiform (see figs 5A and B, 6A). Pereiopod 8 of female with epipodite shorter than the sum of the lengths of basipodite and exopodite; exopodite with 4 setae (fig. 6C).

Recorded from Yugoslavia, Bulgaria, Romania, Hungary, Czechoslovakia, Germany, France and Switzerland.

—   Mandible with 6 teeth (pars incisiva 3, pars molaris 3), (fig. 4D–F);
    uropodal sympodite with 4 spines (fig. 4H)—
    Bathynella stammeri Jakobi

Pereiopod 8 of male with anterior plate rectangular and with a prolongation of the external distal angle, internal face with 3 spines; internal lobe extending beyond distal limit of anterior plate; small lobe in profile bilobed (see figs 5C and D, 6B). Pereiopod 8 of female with epipodite equal to or longer than the sum of the lengths of basipodite and exopodite; exopodite with only 2 terminal setae (fig. 6D).

Recorded from Romania, Czechoslovakia, Germany, Italy and England.

Although *Bathynella* has been found in caves and wells in Britain, greater numbers have been collected from the interstitial habitat of superficial riverine sands and gravels. It is probably widespread in the interstitial of the phreatic or permanent water table. In Britain *Bathynella* is recorded from Devon, Wiltshire, Berkshire, Oxfordshire, Yorkshire, Lancashire, Cumberland, Westmorland and Stirlingshire.

Specimens from southern England and Scotland have not been seen by the reviser but all specimens examined from Yorkshire, Lancashire, Cumberland and Westmorland have been identified as *Bathynella stammeri* Jakobi.

For details of the animals' structure see Calman (1917) and Bartok (1944). Details of the taxonomy of the species discussed are given by Serban (1966a, 1966b) and Husmann (1968). For recent records of *Bathynella* in Britain see Efford (1959), Spooner (1961), Maitland (1962), Gledhill & Driver (1964) and Serban & Gledhill (1965). Noodt (1965) discusses the biogeography of the Syncarida. A very detailed analysis of the morphology and systematics of *Bathynella* is presented by Serban (1972).

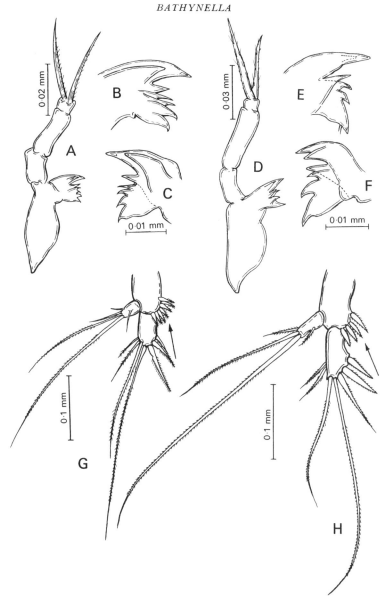

Fig. 4. A, *Bathynella natans* mandible, B and C detail of mandible; D, *B. stammeri* mandible; E and F detail of mandible; G, *B. natans* uropod (⁄ spines on sympodite); H, *B. stammeri* uropod (⁄ spines on sympodite); (A–F after Serban 1966, G and H after Husmann 1968).

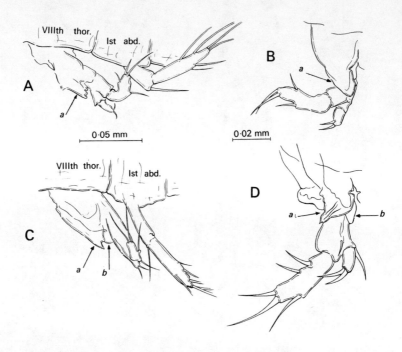

Fig. 5. 8th thoracic segment with male pereiopod 8, and 1st abdominal segment with pleopod 1, of: A, *Bathynella natans*; C, *B. stammeri*. Male pereiopod 8, frontal view, of: B, *B. natans*; D, *B. stammeri*. (Figures after Serban 1966). *a* ↑, anterior plate; *b* ↑, internal lobe.

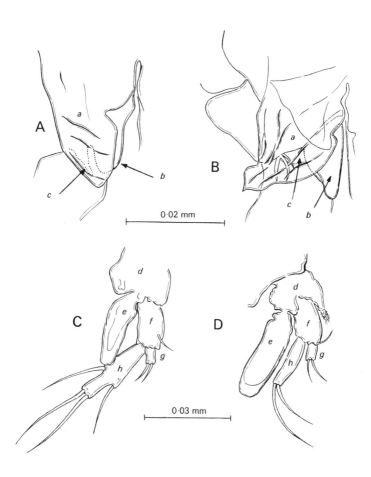

Fig. 6. Basal region of male pereiopod 8 of A, *Bathynella natans*, B, *B. stammeri*; female pereiopod 8 of C, *B. natans*, D, *B. stammeri*. (Figures after Serban 1966). *a*, anterior plate, overlapping *b* and *c*; *b*✶, internal lobe; *c*✶, small lobe; *d*, coxopodite; *e*, epipodite; *f*, basipodite; *g*, endopodite; *h*, exopodite (sympodite = coxopodite + basipodite).

Super-order EUCARIDA Calman, 1904

The Eucarida are characterized by their carapace, which is fused with all the thoracic somites, and by their eyes on stalks. Of the two orders only one, the Decapoda, contains any freshwater species. Allen (1967) gives an illustrated key to all of the British species.

Order DECAPODA Latreille, 1802
Sub-order NATANTIA

To this sub-order belong the familiar prawns and shrimps of the sea shore. Some of the prawns and the common shrimp *Crangon crangon* (L.) are commonly found in brackish water, but it is doubtful whether any in Britain can be truly called freshwater animals by the criterion of breeding. Nevertheless two prawns are very tolerant of fresh water when adult. **Palaemonetes varians** (Leach) (fig. 7) occurs in marsh pools and drainage dykes, some of which are only remotely connected with the sea, or receive sea spray only occasionally. It is recorded from coastal and estuarine localities in south and east England, Lancashire, Anglesey, south Wales and from eight Irish counties; the only records from Scotland are from the Forth estuary. It is absent from the Isle of Man. **Palaemon** (=*Leander*) **longirostris** Milne Edwards is recorded as abundant in Oulton Broad and the River Waveney in Norfolk (Gurney, 1923; Hamond, 1971). Both *P. varians* and *P. longirostris* are almost colourless and translucent when alive. They may be separated by reference to the mandibular palp, which is present in *P. longirostris* but absent in *P. varians* (see Allen, 1967).

The superficially similar Mysidacea belong to the Peracarida and may be distinguished from true prawns by their two-branched thoracic limbs. The so-called "freshwater shrimp" is the amphipod *Gammarus* (fig. 16), whose marine counterparts are the "scuds" (*Marinogammarus* species as well as marine species of *Gammarus*).

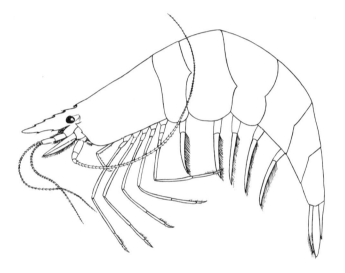

Fig. 7. *Palaemonetes varians*.

Sub-order REPTANTIA
Section BRACHYURA

Strictly no crab should be included here as a freshwater animal, but the Chinese mitten crab, **Eriocheir sinensis** Milne Edwards (fig. 8), seems worth mentioning, since although it breeds in the sea, it lives in rivers and travels such distances up them that it could reach anywhere in Britain. A native of China, it was first noticed in Europe in the River Weser in 1912, having possibly travelled from the east in the ballast tanks of ships. By 1927 it had been seen in rivers up to 400 miles from the sea, had become a pest in Germany because of its habit of burrowing in river banks, and was spreading fast. In 1935 it was known in all the Baltic countries and in the extreme north of France. In the same year a single specimen was taken in the Thames but no more specimens have been reported. Since 1935 it has steadily extended its range round the coasts of France and has reached the Gironde (Hoestlandt, 1955). It is possible that the crab will reach our shores again, and if it then finds its way into our rivers it may well become as numerous as on the Continent. According to De Leersnyder (1967), *Eriocheir* prefers to moult in fresh water but is unable to lay eggs at low salinities.

The crab is easily recognised by the thick mass of hairs on the claws, from which the name is derived (Greek $E\rho\iota o\nu$ = wool and $X\epsilon\hat{\iota}\rho$ = a hand).

Section ASTACURA

The crayfish is like a small lobster (fig. 9). There is only one endemic species in Britain and Ireland—**Austropotamobius pallipes** (Lereboullet) (syn. *Astacus pallipes*, *Potamobius pallipes*) (Gordon, 1963). This is the commonest crayfish in France (Laurent & Suscillon, 1962). *Astacus astacus* (L.) is the only native species in Scandinavia and much of north-western Europe, overlapping with *Austropotamobius torrentium* Schrank in southern Europe and with *Astacus leptodactylus* Eschscholz in eastern Europe (Bott, 1950; Karaman, 1962; Holthuis, 1967; Cukerzis, 1968; Lund, 1969). Some American species of crayfish, notably *Orconectes limosus* (Rafinesque) and recently *Pacifastacus leniusculus* Dana have been introduced into Europe to support the *Astacus astacus* fisheries badly affected by the crayfish plague fungus *Aphanomyces astaci* (Svärdson, 1965, 1972; Unestam, 1969; Abrahamsson, 1972, 1973).

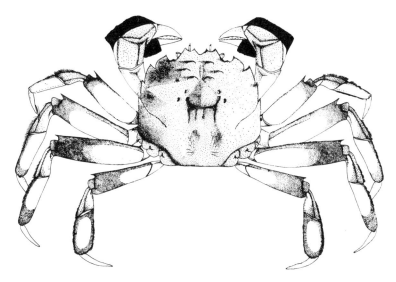

Fig. 8. *Eriocheir sinensis.*

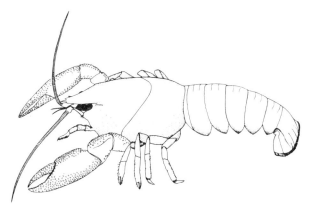

Fig. 9. *Austropotamobius pallipes.*

Since introductions to Britain from Europe have been made in the past, and may well be made again, a key is given to the commonest species found in Europe, based on Bouvier (1940) and Laurent (1960). The nomenclature used is that of Holthuis (1967) who lists four species of *Astacus*, i.e. *astacus* (L.), *colchicus* Kessl., *leptodactylus* Esch. and *pachypus* Rathke (with two subspecies in *A. astacus* and four subspecies in *A. leptodactylus*), and two species of *Austropotamobius*, i.e. *pallipes* (Ler.) and *torrentium* (Schr.) (with four subspecies in *A. pallipes* and two subspecies in *A. torrentium*). Further details may be found in Bott (1950) and Karaman (1962).

1 With lateral spines in front of the cervical groove (fig. 10). Chela smooth except on internal edge. With a spur on the carpopodite of the cheliped— **Orconectes limosus** (Rafinesque) (*Cambarus affinis* Say)

The chela, whose surface is covered in small pits which towards the apex are arranged in parallel rows, feels smooth to the touch whereas the chela of the native European crayfishes feels more or less rough. Note that other crayfishes introduced from America may also have one or more spurs on the carpopodite of the cheliped.

A species from eastern North America introduced into Germany and France at the end of the 19th century to support fisheries decimated by *Aphanomyces astaci*. It is now widely but disjunctly spread in Central Europe.

— No lateral spines in front of the cervical groove. No spur on carpopodite of the cheliped— 2

2 With prominent spines behind the cervical groove (figs 12A, 13A, 14A)— 3

— With insignificant simple tubercles behind the cervical groove (fig. 11A). Rostrum without a dorsal ridge and with the apex in the shape of an equilateral triangle (fig. 11A). Ventral face of scale on antenna 2 has a median ridge (keel) which is denticulate (fig. 11B)— **Austropotamobius torrentium** (Schrank) (*Astacus torrentium* (Schrank))

A small species, seldom more than 8 cm between the rostrum and telson, found in the clear water of mountain streams in Central Europe.

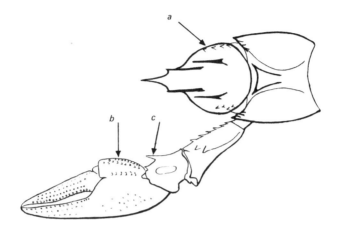

Fig. 10. *Orconectes limosus* (after Laurent 1960): carapace and cheliped in dorsal view. *a* ↗, spines; *b* ↗, tubercles on chela; *c* ↗, spur on carpopodite.

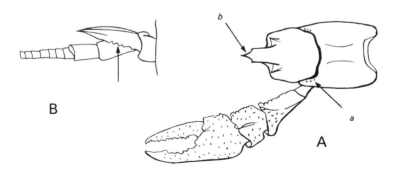

Fig. 11. *Austropotamobius torrentium* (after Laurent 1960): A, carapace and cheliped in dorsal view (*a* ↗, simple tubercles; *b* ↗, apex of rostrum); B, base of antenna 2 in lateral view (↗ denticulate ridge).

3   Rostrum sides almost parallel; anterior median dorsal ridge clearly denticulate (figs 13A, 14A). With two post-orbital ridges— **4**

—   Rostrum sides converging towards the base of the small triangular apex of the rostrum (fig. 12A); median dorsal ridge discrete and not denticulate (fig. 12B). A single post-orbital ridge (fig. 12A). Median ridge on ventral face of scale on antenna 2 not denticulate (fig. 12C). A spur-like projection at the base of the endopodite on pleopod 2 of males (fig. 12D). Three pleural gills—
Austropotamobius pallipes (Lereboullet)
(*Astacus pallipes* Lereboullet)

Widespread in the south and midlands as far north as Northumberland, but apparently absent from Scotland and only one record for Wales (Thomas & Ingle, 1971). Common in Ireland (Moriarty, 1973).

Bott (1950) recognized three subspecies, *A. pallipes pallipes* (Lereboullet), *A. pallipes italicus* (Faxon) and *A. pallipes lusitanicus* (Mateus). Thomas & Ingle (1971) question the distinction between the first two subspecies, but Laurent & Suscillon (1962) examined this problem in some detail and showed that *italicus* has one (rarely two) lateral spines behind the cervical groove, whereas *pallipes* has one to five (average three) lateral spines behind the cervical groove (fig. 12A); also in *italicus* the apex of the rostrum is 1/3 of the total length of the rostrum, whereas in *pallipes* the apex is only 1/5 of the length of the rostrum (x/y, fig. 12A). Using these criteria, specimens from Westmorland are subspecies *pallipes*. Holthuis (1967) lists *A. pallipes italicus* (Faxon) as *A. pallipes fulcisianus* (Ninni).

Karaman (1962) recognized five subspecies, including *A. pallipes carsicus* Karaman (so far known only from Yugoslavia) and *A. pallipes bispinosus* Karaman. Holthuis (1964) showed that the four known specimens of *bispinosus* are actually an American species, *Cambaroides schrenckii* (Kessler). Unfortunately the keys given by both Bott and Karaman are also incorrect with respect to characters for *A. pallipes* (see Gordon (1963) and Thomas (1974)).

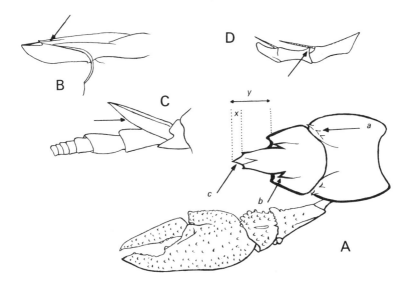

Fig. 12. *Austropotamobius pallipes pallipes* (after Laurent 1960): A, carapace and cheliped in dorsal view (*a* ↗, spines; *b* ↗, post-orbital ridge; *c* ↗, apex of rostrum; *x*, length of apex; *y*, length of rostrum); B, anterior part of rostrum in lateral view (↗ median dorsal ridge); C, base of antenna 2 (↗ median ridge); D, male pleopod 2 (↗ spur).

4 (3) The two post-orbital ridges simple (fig. 13A). Sides of basal part of rostrum smooth; a single line of lateral spines behind the cervical groove (fig. 13A). Telson convex (fig. 13B). Male chela robust, with a hollow between two tubercles on the fixed side (fig. 13A)—
Astacus astacus Linnaeus
(*A. fluviatilis* Fab.)
(*A. nobilis* Huxley)

According to Bott (1950) this species prefers larger streams and rivers with clear water over a muddy bottom. Widely distributed in northern Europe. It has probably been introduced into Britain.

— The second (posterior) post-orbital ridge with spines (fig. 14A). Sides of basal part of the rostrum denticulate; sides of carapace covered with spine-like projections (fig. 14A). Telson slightly concave (fig. 14B). Male chela elongate, biting edge almost straight and with small tubercle (fig. 14A)—
Astacus leptodactylus Eschscholz

An Asian species which is extending its range into western Europe, where it is replacing *Astacus astacus* (Cukerzis, 1968).

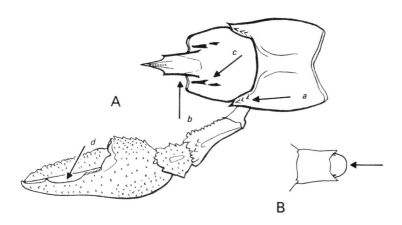

Fig. 13. *Astacus astacus*: A, carapace and cheliped in dorsal view (after Laurent 1960), (*a* ↗, spines; *b* ↗, smooth sides of rostrum; *c* ↗, post-orbital ridge; *d* ↗, hollow on fixed side of male chela); B, telson (after Bouvier 1940), (↗ convex margin).

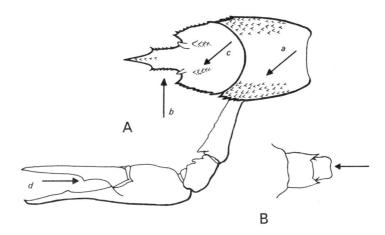

Fig. 14. *Astacus leptodactylus*: A, carapace and cheliped in dorsal view (modified from Bouvier 1940), (*a* ↗, spines; *b* ↗, denticulate rostrum; *c* ↗, spinose posterior post-orbital ridge; *d* ↗, tubercle on biting edge of male chela); B, telson (after Bouvier 1940), (↗ concave margin).

Super-order PERACARIDA Calman, 1905

The characteristic feature of this group of Crustacea is the possession of a brood pouch, in which the young undergo a direct development, usually formed by plates attached to certain of the thoracic limbs. Only five orders, Mysidacea, Amphipoda, Isopoda, Tanaidacea and Spelaeogriphacea contain any freshwater species and of these only the first three occur in the British Isles. The shrimp-like Mysidacea are distinguished from the true decapod shrimps by the facts that, besides possessing a brood pouch, the carapace is fused with only 3 or 4 thoracic segments and the thoracic limbs have two approximately equal branches.

1   Animal with a carapace covering the thoracic (mesosome) segments (fig. 15). Eyes stalked— Order MYSIDACEA, p. 23

—   Animal without a carapace, so that the thoracic segments are visible dorsally. Eyes sessile or absent— 2

2   Body flattened from side to side (fig. 16), or if not flattened the second antenna is very long and held out in front of the animal (fig. 17). Abdominal limbs not flattened, but differentiated into three anterior pairs of pleopods and three posterior pairs of uropods (figs 16 and 17)— Order AMPHIPODA, p. 24

—   Body flattened dorsoventrally (fig. 43). At least some abdominal limbs forming flattened, plate-like gills, which are held against the underside of the abdomen (fig. 45)— Order ISOPODA, p. 58

Order MYSIDACEA Boas, 1833

Only one mysid, **Mysis relicta** Lovén (fig. 15), is a true freshwater species. The body is long and slender and has a characteristic humped appearance when viewed laterally, the body being bent downward in the region of the first and second abdominal segments. The pleopods of the female are rudimentary, unsegmented and setose; on the male the first, second and fifth pleopods are similar to those of the female, the third pair are biramous, the fourth pair are long, extending backwards almost to the posterior end of the telson.

Segerstråle (1957) suggests that *M. relicta* was an arctic estuarine species which became widely distributed during the last ice age and was isolated in fresh water thereafter. On the other hand, Holmquist (1959)

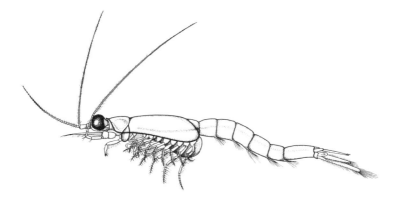

Fig. 15. *Mysis relicta*, female (after Sars 1867).

argues that *M. relicta* was already widespread in fresh water in the northern hemisphere long before the last ice age. In Europe (Fürst, 1965) and Canada (Stringer, 1967) it is transplanted into lakes to provide food for trout. In Britain it has been recorded from Ennerdale Water; in Ireland from Loughs Neagh, Corrib, Ree, Erne and Derg, and from the rivers Shannon and Bann (Tattersall & Tattersall, 1951; Holmquist, 1959).

Neomysis integer (Leach) (= *N. vulgaris* Thompson) is a brackish-water species, which can nevertheless survive long periods of isolation in fresh water. It has never been recorded far from the sea, and when it is found in fresh water there is generally a fairly recent record of the sea's having reached the place.

In *Mysis* the telson (tail fan) is cleft; in *Neomysis* it is entire.

Order AMPHIPODA Latreille, 1816

In the identification of members of this order it is helpful to use various terms which will not be familiar to some people. These can be understood by reference to fig. 16. In some instances it is necessary to know the sex of the individual when using the following key. This is done by examining the ventral side of the thorax (mesosome). In the male the last (8th, but apparently 7th because of fusion of the first with the head) segment bears a pair of papillae between the bases of the legs. In the female some of the anterior legs bear centrally directed flat plates, the oostegites, which form the brood pouch in which the eggs are carried. The oostegites should not, however, be confused with the gills, which are also flat plate-like structures attached to the insides of the legs. Gills are present in both sexes and hang downwards; the oostegites lie inside them and are directed towards the mid-line. The sexes are distinguishable when the animals are about half grown, when, however, both oostegites and male papillae are small. In general males are easier to identify than females. The principal European work on this order is that of Schellenberg (1942), but it does not include all the British freshwater species.

The beginner may confuse the telson (figs 16, 20) with uropod 3, since when viewed from above the small dorsal telson usually lies pressed against the basal segments of the third pair of uropods, with the rami (exopod and endopod) of each uropod extending well beyond the telson. Use a fine needle to lift the telson like a small flap away from the base of the uropods in order to determine the extent of the cleft on the posterior margin. The setae, particularly on the appendages, are diverse in size and structure (Sexton, 1924; Sexton & Spooner, 1940) and are often distinguished as fine "hairs" and thicker "bristles" in the literature on *Gammarus*. Some recent accounts refer to gnathopods 1 and 2 as either legs 1 and 2 or pereiopods 1 and 2, and the five walking legs (fig. 16) then become either legs 3 to 7 or pereiopods 3 to 7.

GAMMARUS

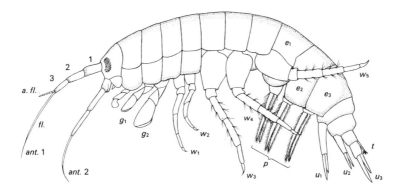

Fig. 16. *Gammarus*: *ant.* 1, *ant.* 2, 1st and 2nd antennae; 1-3, peduncle segments on antenna 1; *fl.*, flagellum; *a. fl.*, accessory flagellum; g_1, g_2, 1st and 2nd gnathopods; w_1-w_5, walking legs; e_1-e_3, epimera; *p*, pleopods; u_1-u_3, uropods; *t*, telson. (Note that there are apparently only seven thoracic segments as the first is fused with the head.)

1   Head and body not markedly flattened.   Antenna 2 very large, more than half as long as the body (fig. 17)—   genus COROPHIUM Latreille

This genus, reviewed by Crawford (1937a) contains 32 marine and brackish-water species and 2 freshwater species. Only one species, **Corophium curvispinum** Sars, of this tubicolous amphipod has so far been recorded from fresh water in Britain. It was originally recorded from the River Avon at Tewkesbury (Crawford, 1935). Moon (1970) recorded *C. curvispinum* in the Grand Union Canal where it is now known to occur from Leicester right through into the middle of Northamptonshire. It is found on *Fontinalis* and algae, silt and sponge debris on the brickwork of locks and bridges. The species has also been found in the River Severn at Stourport, Worcestershire; the Coventry Canal at Lichfield, Staffordshire; the Oxford Canal at Coventry, Warwickshire; the Staffordshire/Worcester Canal near Stafford; and from the Shropshire Union Canal, Cheshire. Another freshwater species, *Corophium spongicolum* Velitchkovsky, has so far not been found.

—   Head and body flattened from side to side.   Antenna 2 less than half the length of the body—   2

2   Antenna 1 very short, equal in length to the first two segments of antenna 2 (fig. 18)—   **Orchestia cavimana** Heller
    (*O. bottae* Milne Edwards)

This species of semi-terrestrial amphipod, previously referred to as *O. bottae* Milne Edwards, was first taken in Cheshire in 1942, since when it has been recorded from several places in the north Midlands, Oxford, Surrey and Norfolk (Fryer, 1950, 1951; Curry, Grayson & Milligan, 1972). Common and widespread in Cheshire (Holland, 1976a).

It is generally found under stones just above the water's edge on the banks of rivers and canals. Like the sand hoppers of the sea shore, to which it is closely related, it escapes in a series of hops when disturbed. A key for the genus *Orchestia* is given by Reid (1947).

—   Antenna 1 and antenna 2 approximately the same length (in most cases antenna 1 is slightly longer than antenna 2)—   3

3   Eyes absent (figs 39, 42)—   4

—   Eyes present (figs 16, 24, 34)—   5

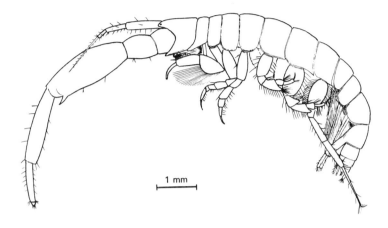

Fig. 17. *Corophium*.

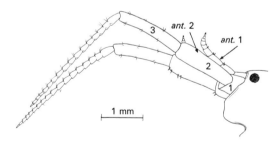

Fig. 18. *Orchestia cavimana*: *ant.* 1, *ant.* 2, 1st and 2nd antennae; 1–3, peduncle segments on antenna 2.

4   Gnathopod hands longer than broad (fig. 19A, B). Telson only shallowly emarginate posteriorly (fig. 19C)—
    Crangonyx subterraneus Bate
    (*Eucrangonyx vejdovskyi* Stebbing) ·

A small subterranean species (up to 6 mm long) recorded from wells, caves and from the interstitial habitat of superficial riverine gravels in southern England and Wales.

—   Gnathopod hands about as broad as long (figs 19D, E and 40). Telson deeply cleft (figs 19F, 38)—
    genera NIPHARGELLUS & NIPHARGUS, **16**, p. 52

5 (3) Dorsal surface of the last three abdominal segments (urosome) without spines or setae (fig. 20A) (dorsal surface of the whole body has a very smooth appearance).   Outer ramus of uropod 3 armed with marginal spines, but setae rare or absent; inner ramus^a of uropod 3 very short, about one fifth the length of the outer ramus (fig. 20C). Telson^a cleft to about the middle (fig. 20B)—
    Crangonyx pseudogracilis Bousfield
(*Eucrangonyx gracilis* (Smith))
(*Crangonyx gracilis* Smith p.p.)

a These characters may not be visible under low magnification. Uropod 3 and the telson should be removed from the animal for examination under high magnification (see p. 24).

A North American species of a world-wide genus, many of whose species are subterranean in habit; recently introduced and evidently still spreading. For discussion of the North American species see Shoemaker (1942) and Bousfield (1958), who has subdivided *C. gracilis* Smith into several species; *C. gracilis* s.str. does not occur in Britain.

C. pseudogracilis inhabits rivers, canals, ponds, lakes and reservoirs, and tolerates saline and polluted water. It has been found in central and southern England and Wales extending north to Yorkshire and Lancashire, with isolated records* from Northumberland, Grangemouth (Stirlingshire), the Lake District and the Norfolk Broads.

---

* Now recorded from a pond in Dublin (Eire)—see Holmes, J. M. C. (1975) *Crangonyx pseudogracilis* Bousfield, a freshwater amphipod new to Ireland. *Ir. Nat. J.* **18**, 225–6.

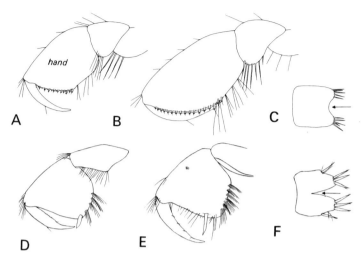

Fig. 19. *Crangonyx subterraneus* (after Schellenberg 1942): A, male gnathopod 1; B, male gnathopod 2; C, female telson (⁄ shallow concavity in posterior margin). *Niphargus aquilex*: D, gnathopod 1; E, gnathopod 2; F, telson (⁄ deep cleft in posterior margin).

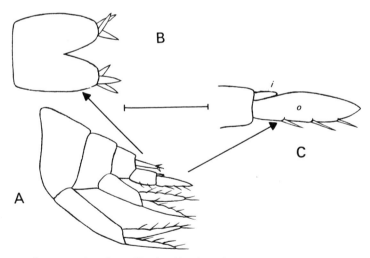

Fig. 20. *Crangonyx pseudogracilis*: A, side view of urosome; B, dorsal view of telson (note cleft in posterior margin); C, dorsal view of terminal segments on uropod 3 (*i*, small inner ramus; *o*, large outer ramus). Scale line: A, 1 mm; B and C, 0·2 mm.

— (5) Dorsal surface of the last three abdominal segments (urosome) armed with spines, or setae, or both (figs 21, 25). Outer ramus of uropod 3 armed with spines and long setae (fig. 22A, B). [b]Telson cleft almost to the base (fig. 22C, D)—   **6**

[b] Four small (3 mm) specimens dredged from *c.* 45 m in Lough Mask and described as *Bathyonyx de Vismesi* (Vejdovsky, 1907) are regarded here as juvenile and possibly aberrant specimens of *Gammarus* (both *G. duebeni* and *G. lacustris* occur in L. Mask).

6   Inner ramus of uropod 3 at least half the length of the outer ramus (except in very small specimens) (fig. 22B). Dorsal surface of each urosome segment with three discrete tufts of spines and setae (fig. 25)—   genus GAMMARUS, **8**

— Inner ramus of uropod 3 less than half the length of the outer ramus (*Marinogammarus finmarchicus* Dahl) or the inner ramus is very short, one fifth or less of the length of the outer ramus (fig. 22A)—   **7**

7   Metasome and urosome segments in mature adult males covered with numerous tufts of long curved setae, sometimes with scattered small spines in between the setae (fig. 21A) (setation much reduced in adult females and juveniles of both sexes (fig. 21B)). Inner ramus of uropod 3 one fifth or less of the length of the outer ramus (fig. 22A)—   **Echinogammarus berilloni** (Catta)

Channel Isles only, in fresh water. Pinkster (1973) gives a key for the *Echinogammarus berilloni* group of species in Europe.

— No setae on metasome. Urosome with setae and spines restricted to three tufts on the dorsal humps of the segments (as in *Gammarus*). Intertidal and marine—   genus MARINOGAMMARUS

A key for this genus is given by Sexton & Spooner (1940).

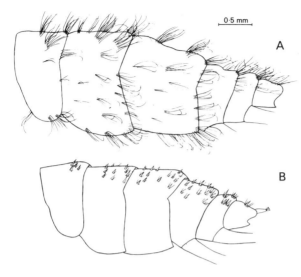

Fig. 21. *Echinogammarus berilloni*: A, hind end of body (metasome and urosome) of A, male; B, female.

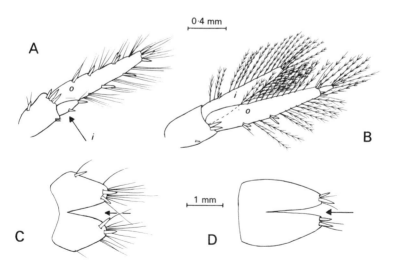

Fig. 22. Uropod 3 of: A, *Echinogammarus berilloni*; B, *Gammarus lacustris* (*i*, inner ramus; *o*, outer ramus). Telson in dorsal view of C, *E. berilloni*; D, *G. lacustris* (↗ deep cleft in posterior margin of telson).

Genus GAMMARUS

Nine species are recorded from fresh and brackish waters in and around the British Isles. In Britain the two common freshwater species are *Gammarus pulex* and *G. lacustris*. Both have rounded eyes. All other members of the genus normally have elongated eyes. *G. duebeni* is the common freshwater species in Ireland. These three species have a marked extension to the lower posterior corner of the basipodite on walking legs 3–5; in *G. pulex* this corner has a slender spine. The posterior corners of epimera 2 and 3 are subrectangular in *G. duebeni* and *G. pulex*; in all other species the corners are acute.

The above three freshwater species are less transparent in life than the remainder. *G. locusta* is marine but is occasionally found in estuaries; it is immediately recognizable by the three triangular urosome segments. In Scotland it may be replaced by *G. oceanicus* which is intermediate in many respects between *G. locusta* and *G. salinus*. More often found in brackish than fresh water are *G. zaddachi*, *G. tigrinus* and *G. chevreuxi*. These three are very "hairy" in appearance, with numerous and sometimes dense tufts of long setae on the antennae, legs, uropods and telson. In mature adult males of *G. tigrinus* and *G. chevreuxi* (occasionally in *G. zaddachi*) many of the long setae are curled.

G. salinus and *G. oceanicus* do not occur in fresh water, except where it trickles over the shore and is subject to tidal influence. These two resemble the less "hairy" specimens of *G. zaddachi*. Both *G. salinus* and *G. zaddachi* are distinguished from all others in the genus by possessing numerous dense tufts of setae on the ventral margin of segment 1 on antenna 1.

Table 2. Distribution of lateral lines or groups of setae on outer face of mandible palp segment 3.

Species	No. of lateral lines
G. pulex	1
G. lacustris	1
*G. duebeni	1
G. chevreuxi	1
G. locusta	1–2
G. tigrinus	2
G. oceanicus	2–3
*G. zaddachi	3–5
*G. salinus	3–5

* Setae also present on ventral margin of mandible palp segment 1.

In *G. zaddachi*, *G. salinus*, *G. oceanicus*, *G. tigrinus* and *G. chevreuxi* the most reliable characters diagnostic for immature as well as adult specimens of both sexes are setation on the mandibular palp and the fifth walking leg. In preserved specimens the mandibular palp is reflexed upwards from the mandible, so that segment 3 of the palp lies between the bases of the second (lower) pair of antennae (fig. 23). The setation is best examined under a binocular microscope on palps removed together with the mandibles so that the *outer* face is readily identifiable. Where possible

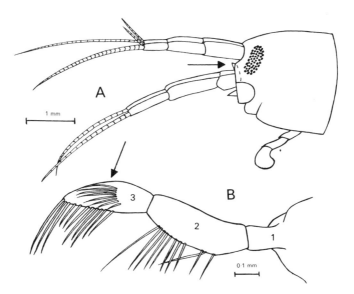

Fig. 23. Head of *Gammarus*: A, showing position of mandible palp (↗); B, detail of mandible palp, with lateral setae shown on segment 3 (↗).

hold the palp in a variety of attitudes to determine the number of lines or groups of setae set obliquely across the outer lateral face of segment 3 (figs 23, 32; Table 2, p. 32). (N.B. Often the inner lateral face of segment 3 has a similar grouping of setae, but the setation here is more variable). For holding the palp, a drop of viscous medium such as lactic acid has the advantage that it will retain the required attitude of the palp long enough to allow inspection under fairly high magnification, and the palp may then be mounted directly into polyvinyl lactophenol on a slide and examined again under a microscope.

8 (6) Eyes short, oval or occasionally kidney-shaped, rarely longer than twice their breadth across the middle (fig. 24A). Walking legs 3, 4, 5 and urosome with spines but few setae (figs 25A, 28A, B). Male* antenna 2 has peduncle segments 3, 4, 5 with sparse, short setae; maximum length of these setae less than twice the width of segment 4 (fig. 27A, B). In side view the dorsal edge of urosome segments more or less straight (fig. 25A). Confined to fresh water—　　9

— Eyes kidney-shaped, at least twice as long as wide in adults and well-grown immature specimens♂ (fig. 24B). Walking legs 3, 4, 5 and urosome with spines and numerous setae (figs 25B, 28C–E). Male* antenna 2 has peduncle segments 3, 4, 5 with numerous tufts of long setae; length of some setae on segment 4 at least equal to twice the width of segment 4 (fig. 27D–F). In side view the dorsal edge of each urosome segment appears either triangular (fig. 36B, C) or "humped" (figs 25B, 36E, F). Fresh and brackish water—　　10

* To distinguish males from females, see p. 24.

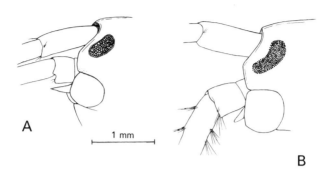

Fig. 24. Heads of: A, *Gammarus pulex*; B, *G. duebeni*.

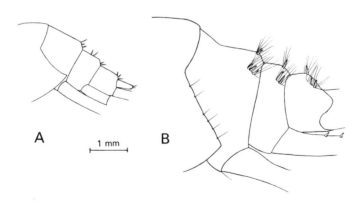

Fig. 25. Urosome of: A, *Gammarus lacustris*; B, *G. duebeni*.

9   Epimeron 2 (side plate of metasome segment 2) has posterior corner of ventral margin not as acutely produced as that on epimeron 3 (fig. 26B). On walking leg 5 (and sometimes on walking legs 3 and 4), the basipodite has a slender spine on the inside of the corner between distal and posterior (ventral) margins; the corner is enlarged posteriorly (fig. 28B). Male antenna 1 relatively long, equal to about half the length of the body. Male antenna 2 flagellum with thick brushes of sensory setae inserted in double rows on the inner margin of all but the most distal segments (setae very dense in the largest individuals); base of flagellum on antenna 2 in male is thick proximally, some segments wider than long (fig. 27B, C). In freshwater streams and lakes, and occasionally in subterranean habitats—
Gammarus pulex (L.)

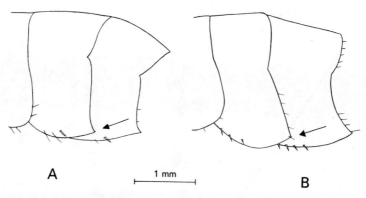

Fig. 26. Epimera of: A, *Gammarus lacustris* (⁄ acute corner); B, *G. pulex* (⁄ rectangular corner).

In waters of moderately high to low mineral content but, in Britain, never where there is any marked influence of brackish water, either from estuaries or mineral springs. (It does occur in these habitats in France and Germany; references in Pinkster, 1972). In Britain it has not been found in water with pH consistently below 5·7. The distribution extends into mountain streams, e.g. numerous in Brownrigg Well near the summit of Helvellyn at 859 m (Sutcliffe & Carrick, 1973a, b). It is absent from some western peninsulas and the extreme north of Scotland (Sutherland and Caithness) and absent from islands, including Orkney, Shetland and the Faeroes, but has recently been successfully introduced into Ireland, into the watershed of the River Bann. It was probably also carried to the Isle of Man by humans (Hynes,

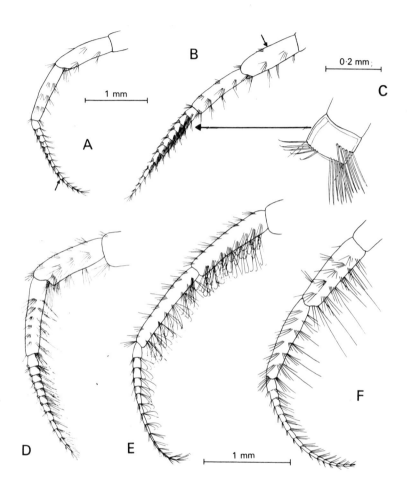

Fig. 27. Antenna 2, peduncle segments 4-5 and flagellum of male: A, *Gammarus lacustris* (♂ calceolus; see also figs. 31, 35); B, *G. pulex* (♂ segment 4); D, *G. duebeni*; E, *G. tigrinus*; F, *G. zaddachi*. C, segment from base of flagellum of male *G. pulex*.

1954a, 1955a, b). Widely distributed in Europe eastwards to Siberia, Turkey and Greece (Pinkster, 1972) but absent from Norway. Thought to be spreading into north-west Europe after the retreat of the last glaciation (Segerstråle, 1954, 1966).

Stock (1969) showed that in conformity with the International Code of Zoological Nomenclature the only valid generic name for species belonging to the *G. pulex* (L.) group is *Gammarus*; all other names such as *Rivulogammarus* Karaman should not be used. Pinkster (1970) published a redescription of *Gammarus pulex* (L.) based on neotype material. The recent resurrection of the species *G. fossarum* Koch, 1835 (Goedmakers, 1972), and the erection of a new species *G. wautieri* Roux (1967), both from within the *G. pulex* complex, leaves only very minor differences between European populations of *G. pulex*. (Pinkster (1972) recognizes three subspecies of *G. pulex*, including *G. pulex pulex* Schellenberg).

— (9) Epimeron 2 has posterior corner of ventral margin as acutely produced as that on epimeron 3 (fig. 26A). On walking legs 3, 4 and 5, the basipodite is without a spine on the distal-posterior corner; the corner is enlarged posteriorly (fig. 28A). Male antenna 1 relatively short, approximately one third the length of the body. Male antenna 2 flagellum without brushes of sensory setae; base of flagellum on antenna 2 in male is less thick proximally, no segment being wider than long (fig. 27A); calceoli fairly prominent (fig. 27A and see fig. 31). In fresh water, particularly in lakes—

Gammarus lacustris Sars

In lakes and their outflows; not in running water except below a lake. Lakes in Scotland, Wales, Ireland and northern England, and one Cheshire mere (Hynes, 1955a, b; Sutcliffe, 1972a). Also widely distributed in lakes in Orkney, Shetland and Faeroes (Sutcliffe, 1974). A northern species occurring across Europe and America. Absent from acid lakes with pH below 6·0 in Norway (Ökland, 1969); similarly not recorded from acid lakes in Britain and Ireland. In some lakes occurs with *G. pulex* (Britain) or *G. duebeni* (Ireland) but usually a lake contains only one species of *Gammarus*.

10 (8) Walking leg 5 has basipodite with two[c] spines on the corner between distal and posterior (ventral) margins; distal margin of basipodite more or less the same width as proximal margin of the ischiopodite (fig. 28D, E). Posterior corners of epimera 2 and 3 acute (fig. 29B, C)— 11

[c] The inner spine of this pair is sometimes missing, especially in *G. zaddachi* where the spines are rather delicate and where the inner spine also sometimes resembles one of the stronger setae in the tuft arising from the distal-posterior corner of the basipodite.

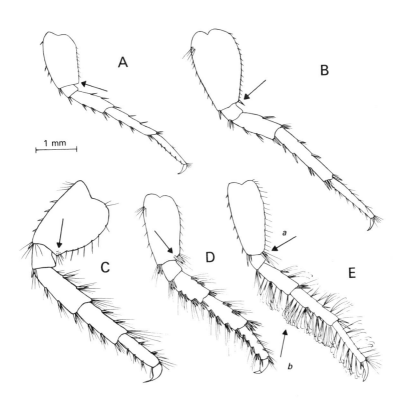

Fig. 28. Walking leg 5 of male: A, *Gammarus lacustris* (↗ extended corner of basipodite, lacking a spine); B, *G. pulex* (↗ spine on inside of extended corner); C, *G. duebeni* (↗ extended corner, lacking spines); D, *G. zaddachi* (↗ spines on posterior corner of basipodite); E, *G. tigrinus* (a ↗, spines; b ↗, curled setae).

— (10) Walking leg 5 has basipodite without spines on the corner between distal and posterior (ventral) margins (a tuft of setae arises from the inner face of the basipodite at this corner); the corner is enlarged posteriorly so that the distal margin of the basipodite is prominently wider than the proximal margin of the next segment[d] (ischiopodite) (fig. 28C). Posterior corners of epimera 2 and 3 subrectangular (fig. 29A) (setae on posterior margins set in small notches). Mandible palp segment 3 with one oblique line of 4–7 lateral setae on outer face (fig. 32A) and usually two oblique lines or groups of lateral setae on the inner face of segment 3. Ventral margin of mandible palp segment 1 with at least one seta near the distal end (fig. 32A)—

Gammarus duebeni Liljeborg[e]

[d] This character is even more pronounced on walking legs 3 and 4.

[e] *G. duebeni* was described by Vilhelm Liljeborg in 1852. Around 1860 he changed his name to William Lilljeborg.

Brackish water; trickles and pools on edge of rocky shores etc. but common in freshwater streams and lakes in Ireland, southern tip of the Isle of Man, some Western Isles (Scotland), Orkney and Shetland. Also occurs in fresh water on some western peninsulas on mainland Britain (Lizard, Cornwall; Holy Island, north Wales; Stranraer, and Kintyre, Argyll) (Hynes 1954a; Sutcliffe 1967, 1974).

That populations of *G. duebeni* in fresh water may belong to a physiological race distinct from populations in coastal brackish waters has been discussed by several workers (Reid, 1939; Hynes, 1954a; Sutcliffe & Shaw, 1968; Pinkster et al., 1970). Sutcliffe (1970, 1971a, b) subsequently showed that specimens from freshwater localities in Ireland are physiologically distinguishable from animals living in both fresh and brackish water in western Britain, but the differences are phenotypic and determined by habitat salinity.

Two subspecies of *G. duebeni* were recognized by Stock & Pinkster (1970) and Pinkster et al. (1970), based on the ratio of length/width of the meropodite on walking leg 5. Allometric growth affects this ratio, but males of *G. duebeni duebeni* are distinguished from males of *G. duebeni celticus* by another ratio, $\bar{Y}/\bar{X}$ or mean $\log_{10}$ meropodite width/mean $\log_{10}$ meropodite length on walking leg 5. Width and length are expressed as mm $\times$ 10 converted to logarithms. The means of the logarithmic values (Y and X) are used to calculate the ratio, which normally lies between 0.71 and 0.73 in populations supposed to be *celticus*, and 0.74 and 0.77 in populations supposed to be *duebeni* (Sutcliffe, 1972b). There is a similar difference in the ratio $\bar{Y}/\bar{X}$ for the carpopodite on walking leg 5, with values of 0.51–0.55 in *celticus* and 0.59–0.61 in *duebeni*. Single specimens cannot be assigned to a subspecies; ideally a sample of at least 30 males is required to determine $\bar{Y}/\bar{X}$. *G. duebeni celticus* occurs throughout Ireland, in both fresh and brackish habitats, also on the Isle of Man and the Shetland Isles. *G. duebeni duebeni* is characteristic of brackish habitats on the east and west

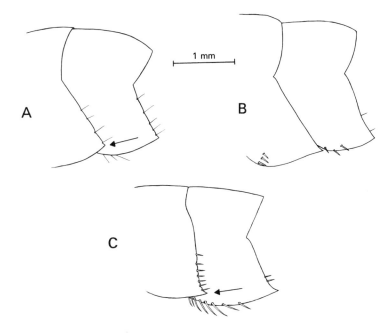

Fig. 29. Epimera 2 and 3 of: A, *Gammarus duebeni* (⤴ sub-rectangular corner); B, *G. tigrinus*; C, *G. zaddachi* (after Sexton 1942), (⤴ acute corner).

coasts of Britain. Populations of *G. duebeni* living in freshwater streams on the Lizard, Stranraer and Kintyre peninsulas are morphologically intermediate between the two but, on the basis of $\bar{Y}/\bar{X}$ for the meropodite on walking leg 5, the Stranraer population is *celticus*, the Lizard[f] and Kintyre populations are *duebeni*.

[f] Due to an error in $\bar{X}$, the Lizard population was incorrectly assigned to *celticus* by Sutcliffe (1972b, Table 7). The correct value for $\bar{X}$ in males is 0·936, and $\bar{Y}/\bar{X}$ = 0·743 (in females, $\bar{Y}/\bar{X}$ = 0·747). In another sample obtained in May 1973, meropodite $\bar{Y}/\bar{X}$ = 0·76 in both males and females.

11 Accessory flagellum of adult male antenna 1 shorter than peduncle segment 1 (fig. 30C–E). Ventral margin of peduncle segment 1 on antenna 1 with three to six strong tufts of setae (fig. 30C–E) or, if setation sparse here, then long fine setae (often curled in adult males) dense on some or all of walking legs 1–5. Generally confined to brackish water or freshwater habitats with some intermittent connection with the sea— 12

— Accessory flagellum of adult male antenna 1 approximately equal in length to or longer than peduncle segment 1; ventral margin of peduncle segment 1 with sparse setae (fig. 30A, B). Setae short and very sparse on walking legs 1–5. Marine littoral and estuarine, rarely found where fresh water trickles over the shore— 15

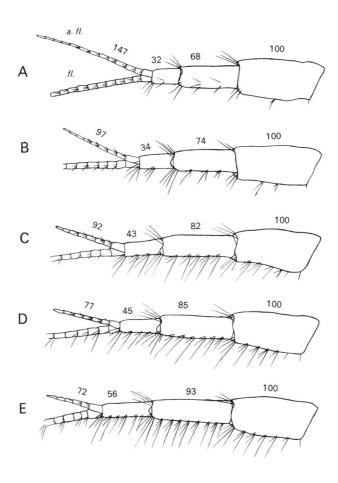

Fig. 30. Antenna 1 of male: A, *Gammarus locusta*: B, *G. oceanicus*; C, *G. salinus*; D, E, *G. zaddachi* (after Spooner 1951). The numbers represent the relative lengths of peduncle segments 2 and 3, and the accessory flagellum, expressed as a percentage of the length of peduncle segment 1.

12 Mandible palp segment 1 with one or more setae at distal end of ventral margin, segment 3 with three to five oblique lines or groups of lateral setae on outer face (fig. 32C). Antenna 1 has peduncle segment 1 with three to six dense tufts of setae[g] (fig. 30C–E). Curled setae not present (except rarely on antenna 2 in adult male). Adult male, and some females, with calceoli on proximal segments of flagellum of antenna 2 (fig. 31)— 13

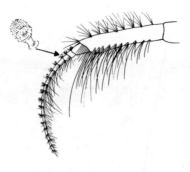

Fig. 31. Flagellum of antenna 2 of male *Gammarus zaddachi* showing calceoli (↗ calceolus) (after Sexton 1942).

— Mandible palp segment 1 without setae on ventral margin, segment 3 with one or two oblique lines or groups of lateral setae on outer face (fig. 32B). Antenna 1 has peduncle segment 1 with few setae, usually 1-3 single setae or sparse tufts of 2–3 setae on ventral margin[g]. Adult males with numerous long setae, some prominently curled, in dense tufts on antenna 2, maxillipeds, gnathopods 1 and 2, and at least on walking leg 1. Flagellum of antenna 2 without calceoli— 14

[g] Excluding the distal terminal tuft of setae present on the ventral margin of peduncle segment 1.

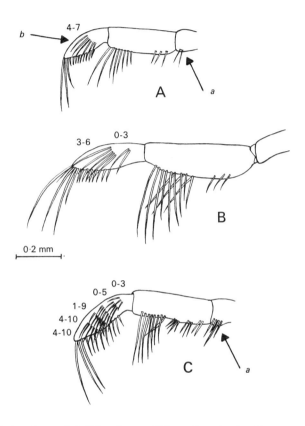

Fig. 32. Outer face of left-hand mandible palp of: A, *Gammarus duebeni*; B, *G. tigrinus*; C, *G. zaddachi*. *a ↗*, ventral setae on segment 1; *b ↗*, lateral setae on segment 3 (the numbers refer to range of setae in each group). (After Kinne, 1954, and Nijssen & Stock, 1966).

13A Mature adult specimens only; walking leg 5, posterior margin of carpopodite (and often of meropodite) with setae longer than spines, or longest seta > longest spine (fig. 33A, B). Normally with numerous dense tufts of setae on antennae and walking legs 1–5—
Gammarus zaddachi Sexton

13B Mature adult specimens only; walking leg 5, posterior margin of carpopodite (and often of meropodite) with setae shorter than spines (fig. 33C). Setation noticeably less dense than in typical *G. zaddachi*, especially on walking legs 1–5— **Gammarus salinus** Spooner

13C Juveniles of *G. zaddachi* often have the characters of 13B. Most juveniles[h], and all adults, are distinguished by the ratio W/Sm on walking leg 5, where W is the maximum width of the basipodite, and Sm is the mean length of the three longest setae on the posterior margin of the basipodite. Measurements should be made with a micrometer eyepiece (Dennert et al. 1969).
W/Sm less than 6·5— **G. zaddachi**
W/Sm greater than 10·0— **G. salinus**

[h] Rygg (1974) gives a key separating juveniles (body length 1·8–4·5 mm) of *G. zaddachi*, *G. salinus*, *G. duebeni*, *G. oceanicus* and *G. locusta*.

For further details on these two species see Sexton (1942), Spooner (1947, 1951), Segerstråle (1947), Kinne (1954), Dennert et al. (1969). Both are found in brackish water round the coasts of Britain and Ireland. *G. zaddachi* can penetrate into freshwater streams, but apparently only breeds in places where there is at least an occasional incursion of sea water. Specimens kept in flowing fresh water from Windermere (Sutcliffe, 1970) lived for up to six months but did not establish a breeding population. Early records of *G. zaddachi* from freshwater loughs in Ireland were incorrect (Hynes, 1951).

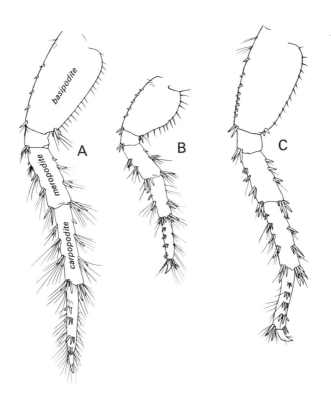

Fig. 33. Walking leg 5 of: A, male and B, female *Gammarus zaddachi*; C, male *G. salinus* (after Spooner 1947).

14 (12) Adult male[i] antenna 1 longer than antenna 2. Adult male with dense tufts of long setae, some curled, on antenna 2, gnathopods 1 and 2, walking leg 1 and uropod 3. Walking legs 2–5 with spines, but few setae (fig. 34). Mandible palp segment 3 usually with one oblique line or group of lateral setae on outer face. Both antenna 1 and antenna 2 without calceoli— **Gammarus chevreuxi** Sexton

[i] N.B.—The characters given here for *G. chevreuxi* and *G. tigrinus* (see below) refer to mature males only. In some female specimens of *G. tigrinus*, antenna 1 is longer than antenna 2, mandible palp segment 3 may have only one oblique line of lateral setae, and setation on the legs of immature specimens may resemble that of *G. chevreuxi*.

Brackish water in Devon, Cornwall, N. Wales and N.E. Ireland (Crawford, 1937b; Spooner, 1957). Described by Sexton (1913, 1924) and the subject of many studies on growth, and pigmentation of the eye (Sexton & Clark, 1937).

— Adult male[i] antenna 1 shorter than antenna 2. Adult male with dense tufts of long setae on all appendages except the telson and antenna 1, with long curled[j] setae prominent on antenna 2 and on the meropodite and carpopodite of walking legs 1, 2, 4 and 5 (figs 28E, 35B). Mandible palp segment 3 usually with two oblique lines or groups of lateral setae on outer face (fig. 32B). Minute, slender calceoli on all but first and last segments of flagellum on male antenna 1 (fig. 35A). No calceoli on antenna 2. Colour in life yellowish with very distinct dark bands— **Gammarus tigrinus** Sexton

[j] Curled setae may be absent in younger mature males and in winter specimens.

Hynes (1954b) showed that what had been known in Britain as *G. tigrinus* Sexton, 1939, was identical with the common American species *G. fasciatus* Say. Bousfield (1958), however, has shown that *G. fasciatus* Say includes a freshwater species for which the name is retained and a brackish-water species for which he revives the name *G. tigrinus* Sexton, since it is this species which is present in Britain. Introduced into Germany in 1957 and now spreading in slightly brackish water (Fries & Tesch, 1965; Nijssen & Stock, 1966; Chambers, 1973).

In coastal brackish waters, and in fresh waters with a raised ion content in the north and west Midlands (see Holland, 1976b). Also abundant in one freshwater locality (L. Neagh, N. Ireland) with a relatively low ion content (Hynes, 1955a, b; Sutcliffe, 1968).

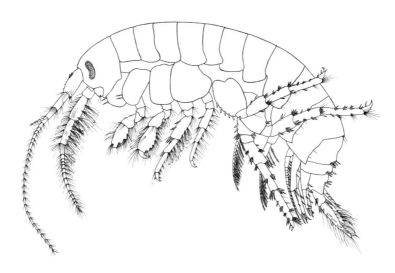

Fig. 34. Adult male (second stage) *Gammarus chevreuxi* (after Sexton 1924).

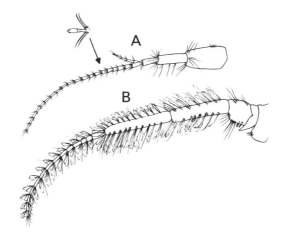

Fig. 35. Antennae of male *Gammarus tigrinus* (after Sexton 1939): A, antenna 1, with slender calceoli on ventral surface of flagellum (⚹ calceolus); B, antenna 2.

15 (11) Dorsal edge of urosome segments 1 and 2 markedly elevated and triangular, with the elevations angled at about 90° (fig. 36B, C). Head lobe in adult sloping forward, upper angle (in front of eye) somewhat acute (fig. 36A). Male antenna 1 has accessory flagellum usually longer than peduncle segment 1 which is approximately equal to the combined lengths of peduncle segments 2 and 3 (fig. 30A); ventral margin of peduncle segment 1 usually with one sparse (0–2) tuft of setae (excluding the distal terminal tuft of setae present on the ventral margins of segments 1–3 on antenna 1), segment 2 with one or two tufts, segment 3 with one sparse tuft or often no setae (fig. 30A). Mandible palp segment 1 without setae on ventral margin, segment 3 with one or two oblique lines or groups of lateral setae on outer face— **Gammarus locusta** (L.)

The forward-sloping head lobe and acute upper angle distinguish *G. locusta* from other members of the genus.

Marine coastal and intertidal zone of shore, also at mouths of estuaries; rarely found in less saline waters with *G. salinus*. Distributed round the coasts of Britain north to Scotland, where it may be replaced by *G. oceanicus* (Spooner, 1951). Spooner (1947, p. 15–16) notes three variant populations of *G. locusta* from the south coast of England; Stock (1967) regards these as belonging to the European species *G. crinicornis* Stock (population at Whitsand Bay, Cornwall) and the sibling species *G. insensibilis* Stock and *G. inaequicauda* Stock (populations in The Fleet, Dorset, and New England Creek, Essex).

— Dorsal edge of urosome segments 1 and 2 less markedly elevated, and angled at approximately 135° (fig. 36E, F). Head lobe in adult not sloping forward, upper angle obtuse (fig. 36D). Male antenna 1 has accessory flagellum slightly shorter than peduncle segment 1 which is slightly shorter than the combined lengths of peduncle segments 2 and 3 (fig. 30B); ventral margin of peduncle segment 1 with two or rarely three sparse tufts of setae, segment 2 with usually three tufts, segment 3 with one strong tuft of setae (fig. 30B). Mandible palp segment 1 usually without setae on ventral margin (occasionally 1–2 setae present), segment 3 with two or three oblique lines or groups of lateral setae on outer face— **Gammarus oceanicus** Segerstråle

Marine coastal and intertidal zone of shore, also in estuaries with *G. salinus*, *G. zaddachi* and *G. duebeni*. A northern species which may replace *G. locusta* in Scotland, extending down to north Yorkshire on the east coast of Britain (Spooner, 1951). Rygg (1974) gives a key separating juvenile specimens of *G. oceanicus* from *G. locusta*.

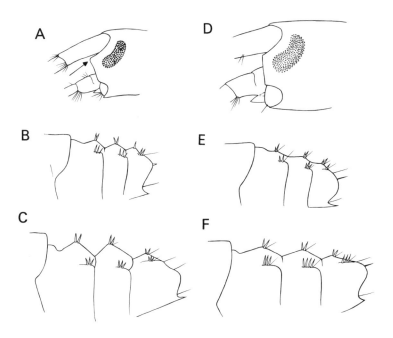

Fig. 36. Head and urosome of: A–C, *Gammarus locusta* (↗ upper angle of head lobe); D–F, *G. oceanicus*; B and E from immature stages, the remainder from adult males (after Spooner 1951).

Genera NIPHARGELLUS and NIPHARGUS

16 (4, p. 28) Mandible palp segment 3 with some terminal setae, but completely lacking the setal fringe (fig. 37A). Telson lobes devoid of spines (but with 3 distal setae of which 2 are long and flexible) (fig. 38A). Gnathopods as in fig. 40C. Uropods 1 and 2 with the outer ramus distinctly shorter than the inner, and lacking lateral spines (fig. 37C, D); uropod 3 short, less than three times the length of the telson. Small species, length *c.* 3 mm (fig. 39)—
Niphargellus glenniei (Spooner)

Originally placed in the genus *Niphargus* this subterranean amphipod is an inhabitant of interstitial groundwater and is recorded from caves, wells, springs and river gravels in Devon.

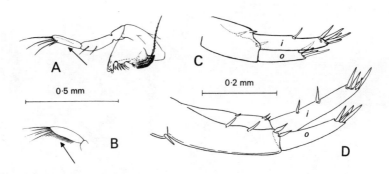

Fig. 37. A, mandible of *Niphargellus glenniei* (⁄ setal fringe absent); B, mandible palp segment 3 of *Niphargus aquilex* (⁄ setal fringe); C, uropod 2 inner view and D, uropod 1 outer view, both of *Niphargellus glenniei*. *i*, inner ramus; *o*, outer ramus.

— Mandible palp segment 3 bearing a fringe of setae (fig. 37B) (this setal fringe may be reduced in immature specimens). Telson with spines, at least a distal group (fig. 38B, C). Gnathopods as in fig. 40A, B, D, E. Uropods 1 and 2 with the rami equal in length or the outer ramus slightly shorter than the inner, the outer bearing one or more lateral spines which are usually smaller than those on the inner ramus; uropod 3 at least three times the length of the telson, usually more, sometimes much more. Larger species normally attaining 5–15 mm— genus NIPHARGUS, **17**

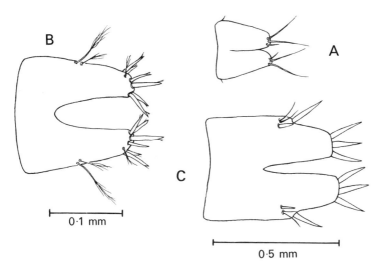

Fig. 38. Telsons of: A, *Niphargellus glenniei*; B, *Niphargus kochianus kochianus*; C, *Niphargus fontanus*.

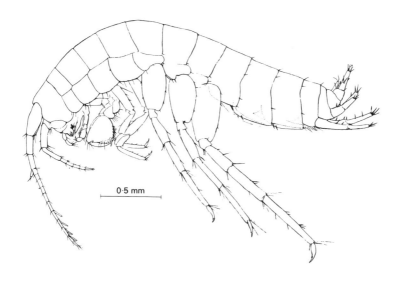

Fig. 39. *Niphargellus glenniei*, female (after Spooner 1952).

54   AMPHIPODA

17 Palmar angle of hands of gnathopods 1 and 2 obtuse (though approaching rectangular); hand of gnathopod 2 of both sexes longer than the carpopodite (fig. 40A, B, E). Uropods 1 and 2 with rami equal in length. Epimeron 3 (side plate of 3rd metasomal segment) posterior angle, even if much rounded off, possessing a spinule (fig. 41A, B). Telson lobes with two or three distal spines and one or more lateral or dorsal spines (fig. 38C).   Male uropod 3 enlarged (fig. 42)—   19

— Palmar angle of hands of gnathopods 1 and 2 subacute; hand of gnathopod 2 in female not longer than the carpopodite (fig. 40D). Uropods 1 and 2 with the outer ramus slightly shorter than the inner. Epimeron 3 posterior angle without a spinule. Telson lobes with 3 or more distal spines. In both sexes, carpopodite of gnathopod 2 with characteristic groups of setae on inferior margin—
Niphargus kochianus s.l., 18

18 Hand of gnathopod 1 roughly square in outline, hand of gnathopod 2 rectangular, palmar margin slightly convex. Telson invagination wide, distally about as wide as the distal width of a telson lobe; the inner margin of each lobe slightly concave (fig. 38B). Epimeron 3 posterior angle distinctly acute. Urosome segment 2 with a single spine dorsally on each side. Female with carpopodite of gnathopod 2 elongate, longer than hand (fig. 40D). Male[k] with carpopodite of gnathopod 2 elongate but not longer than hand—
Niphargus kochianus kochianus Bate

[k] The discovery of males, differing from the description given by Schellenberg (1942) of male *N. kochianus kochianus* from Belgium, together with typical females from near the type locality (Ringwood, Hampshire), suggests that they are males of the nominate subspecies and that the Belgian and other continental material previously considered as that subspecies should be regarded as different, at least at the subspecific level (Stock & Gledhill, in press).

Recorded from caves, wells, springs and interstitial groundwater in southern England.

Fig. 40. Gnathopods of: A, *Niphargus aquilex* (gnathopod 1); first (1) and second (2) gnathopod hands of: B, *N. fontanus*; C, *Niphargellus glenniei*, female; D, *Niphargus kochianus kochianus*, female; E, *Niphargus aquilex*. (⁁ palmar angle). Scale lines 0·2 mm.

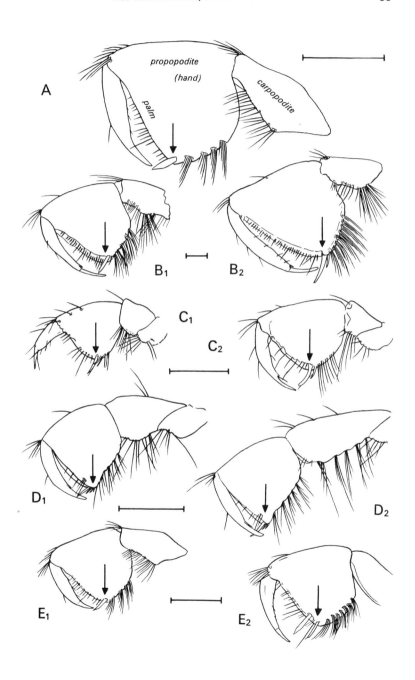

— (18) Gnathopod hands 1 and 2 in area of palmar angle distinctly produced so that hands, especially in female, are less quadrate in outline; palmar margin strongly convex. Telson invagination narrow, the inner margin of the lobes convex. Epimeron 3 posterior angle subrectangular. Urosome segment 2 with two or more spines dorsally on each side. Female with carpopodite of gnathopod 2 elongate, longer than hand. Male with carpopodite of gnathopod 2 elongate, not longer than hand; palmar angle of gnathopods 1 and 2 rounded— **Niphargus kochianus irlandicus** Schellenberg

Lake bottoms (Lough Mask) and subterranean waters in Ireland (Hazelton, 1974a, b, c).

19 (17) Epimera 2 and 3 with the posterior angle much rounded (fig. 41A). Hands of gnathopods 1 and 2 of about equal size (fig. 40E). Uropod 3 of mature adult male has outer ramus with segment 2 about as long as segment 1 (fig. 42); uropod 3 of female has outer ramus with segment 2 one third to one half as long as segment 1—
Niphargus aquilex Schiödte

Schellenberg (1942) recognized three subspecies, *Niphargus aquilex aquilex* Schiödte (recorded from Britain), *N. aquilex schellenbergi* Karaman and *N. aquilex vejdovskyi* Wrzesniowski. Recently Ronneberger and Straskraba in Straskraba (1972) statistically separated as independent species *Niphargus aquilex* Schiödte and *Niphargus schellenbergi* Karaman. The subspecies *N. aquilex vejdovskyi* represents only old specimens of *N. aquilex* Schiödte (Straskraba, 1972) and should therefore be considered synonymous with the latter species.

Wells and other subterranean waters in southern England and Wales. Apparently widely distributed in ground water and liable to be found where this reaches the surface, e.g. springs. Often abundant in the interstitial water of superficial riverine gravels (Gledhill & Ladle, 1969). Observations on the life-history of this amphipod were made by Gledhill and Ladle (op. cit.).

— Epimera 2 and 3 with the posterior angle subrectangular (fig. 41B). Gnathopod 2 hand clearly larger (longer and wider) than that of gnathopod 1 (fig. 40B). Uropod 3 of mature adult male has outer ramus with segment 2 about half as long as segment 1; uropod 3 of female has outer ramus with segment 2 about one third as long as segment 1— **Niphargus fontanus** Bate

Generally in pools in caves, southern England and Wales; also obtained from wells and from riverine gravels but in the latter not so common as *N. aquilex*.

The distribution of the British species of subterranean amphipods is discussed by Glennie (1967, 1968).

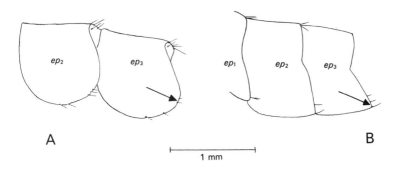

Fig. 41. Second and third epimera of A, *Niphargus aquilex*; B, *N. fontanus* (↗ spinule on epimeron 3).

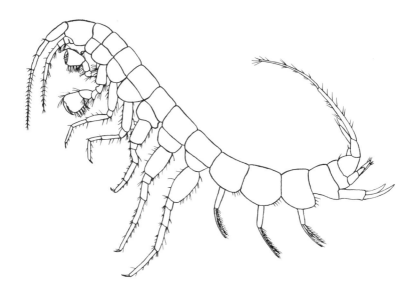

Fig. 42. *Niphargus aquilex*, male (after Chevreux and Fage in Schellenberg 1942).

Order ISOPODA Latreille, 1817

Only one genus*, *Asellus* (fig. 43A), is a true inhabitant of British fresh waters, in that it breeds there, and only four species have so far been recorded from the British Isles (Racovitza, 1919; Tattersall, 1930; Moon,

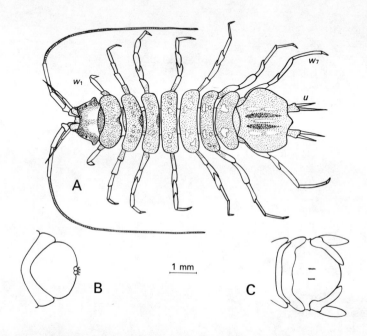

Fig. 43. A, *Asellus* in semi-diagrammatic dorsal view. w_1 and w_7, 1st and 7th walking legs; u, uropod. Dorsal views of the hind ends of B, *Jaera nordmanni* and C, *Sphaeroma hookeri*. (Note that as in the Amphipoda (fig. 16) there are apparently only seven thoracic segments because the first is fused with the head).

1953: Williams, 1972). Two other genera may, however, be met with in dilute brackish water apparently not in direct communication with the sea. These are *Jaera* and *Sphaeroma*, which are readily separable from one another and from *Asellus* by the appearance of the abdomen in dorsal

* *Jaera nordmanni* apparently breeds in L. Tingwall, but this may be related to the relatively high sodium chloride content of the water (1-2 mM).

view (fig. 43). The last pair of appendages (uropods*, see figs 1, 16 and 43 for terminology) of *Jaera* are inconspicuous, those of *Sphaeroma* are broad and plate-like, and those of *Asellus* are elongate. The brackish-water species most likely to be encountered by the freshwater investigator are **Jaera nordmanni** (Rathke), **Sphaeroma hookeri** (Leach) and **S. rugicauda** (Leach). The last two are readily separated by the presence of two small ridges on the top of the abdomen in *S. hookeri* (shown as two black lines in fig. 43C) and their absence in *S. rugicauda*. *J. nordmanni* is locally common, particularly on south and west coasts, under stones in freshwater streams flowing over the shore; it also occurs in fresh water in L. Tingwall on the mainland of Shetland. Further details and keys for the genera *Jaera* and *Sphaeroma* may be found in Naylor (1972).

Genus ASELLUS

Although non-sexual characters may be used to distinguish the British species, the most satisfactory characters are sexual. The sexes differ mainly in the pleopods (abdominal appendages*) and, to a lesser extent, in the first (and fourth) walking legs. These appendages therefore supply the most useful specific characters. In the female the first visible pair of pleopods is morphologically the second, as the first pair is missing; they are of simple construction and much smaller than the following pairs. In the male all the pleopods are retained, but the first two pairs are small and are modified for use in copulation. This difference between the sexes is indicated in fig. 45.

In using the following key it should be noted that the sexes are distinguishable at a length of about 3·5 mm and that the pattern of head pigmentation in all three pigmented species (first noted as different in *A. aquaticus* and *A. meridianus* by Scourfield, 1940) should be used with caution as it displays some variation. Moreover, unpigmented specimens of the pigmented species may occur rarely or in certain situations or after being preserved in certain sorts of preserving fluids for long periods.

* Note that in the Amphipoda the first three pairs of appendages of the abdomen are called pleopods and the last three uropods, whereas in the Isopoda the first five are pleopods and only the last pair are called uropods. Similarly, there being no gnathopods in the Isopoda, all seven pairs of large thoracic (mesosome) appendages are called walking legs (Table 1, p. 6).

1   Pigmented and with eyes (fig. 44B, C)— 2

—   Pigment and eyes absent (fig. 44A; see also figs 47A, 48A, 46B)—
 Asellus cavaticus Schiödte

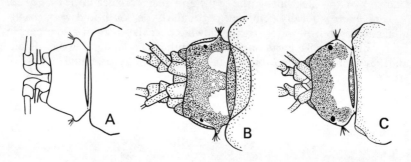

Fig. 44. Head patterns, dorsal views of heads of: A, *Asellus cavaticus*; B, *A. aquaticus* and *A. communis*; C, *A. meridianus*.

2   Possessing a pair of backward-pointing copulatory styles on the ventral surface at the hind end of the thorax (fig. 45B). The first two pairs of pleopods much smaller and obviously different from those following; the second pair of pleopods complex in structure (fig. 47B–D)— MALES, 3

—   Not possessing a pair of backward-pointing copulatory styles on the ventral surface at the hind end of the thorax. Only the first pair of pleopods is different from the following pairs and is of simple construction (fig. 48B, C)— FEMALES, 5

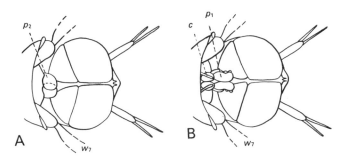

Fig. 45. *Asellus aquaticus*, hind ends of A, female and B, male, in ventral view. w_7, base of walking leg 7; c, copulatory style; p_1, pleopod 1 of male (with pleopod 2 lying beneath it); p_2, "first" pleopod of female (actually equivalent to pleopod 2 in the male, pleopod 1 being absent in the female).

MALES

3   Pigment pattern on the dorsal surface of the head usually as in fig. 44B. The distal segment of pleopod 1 possessing a shallow notch on its outer edge and numerous long setose spines on its distal and outer-distal edges (fig. 47B$_1$). The inner distal segment of pleopod 2 with a spur projecting downwards from the inner side of its base (fig. 47B$_2$). Penultimate segment of the first pair of walking legs possessing a large triangular projection near the end of its distal edge (fig. 46A)—   **Asellus aquaticus (L.)**

—   Pigment pattern on the dorsal surface of the head usually as in fig. 44B or C. The distal segment of pleopod 1 without a conspicuous shallow notch on its outer edge, and distal and outer-distal edges with numerous short simple spines (fig. 47C$_1$, D$_1$). The inner distal segment of pleopod 2 lacks a spur projecting downward from the inner side of its base, although a blunt protuberance may occur here (fig. 47C$_2$, D$_2$). Penultimate segment of the first pair of walking legs possessing either a large triangular projection near the *middle* of its distal edge or no projection at all (fig. 46C, D)—   4

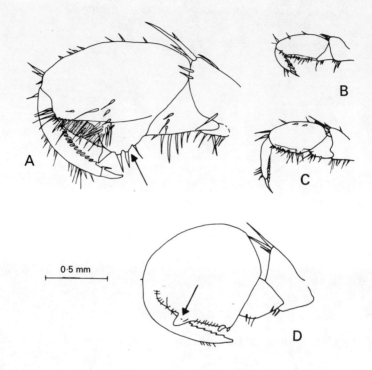

Fig. 46. Tips of walking leg 1 in males of: A, *Asellus aquaticus* (↗ triangular projection); B, *A. cavaticus*; C, *A. meridianus*; D, *A. communis*. The triangular projections (↗) on the distal edge of the penultimate segments in *A. aquaticus* and *A. communis* are less pronounced in small specimens.

4   Pigment pattern on the dorsal surface of the head usually as in fig. 44C (sometimes this pattern is masked by the underlying darker gut contents). Pleopod 1 as in fig. $47C_1$. Outer distal segment of pleopod 2 about the same length as the basal segment; inner distal segment without basal protuberances and with an umbrella-like projection at its tip (fig. $47C_2$). Penultimate segment of the first pair of walking legs without triangular projection on distal edge (fig. 46C)—   **Asellus meridianus** Racovitza

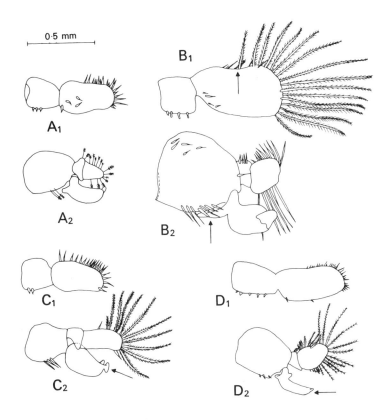

Fig. 47. Left pleopods in ventral view, 1st above, 2nd below, of males of: A, *Asellus cavaticus*; B, *A. aquaticus* (B_1 ↗, shallow notch; B_2 ↗, spur); C, *A. meridianus* (C_2 ↗, umbrella-like projection); D, *A. communis* (D_2 ↗, tube-like projection).

— (4) Pigment pattern on the dorsal surface of the head usually as in fig. 44B. Pleopod 1 as in fig. $47D_1$. Outer distal segment of pleopod 2 shorter than basal segment; inner distal segment with inner and outer basal protuberances and with a simple tube-like projection at its tip (fig. $47D_2$). Penultimate segment of the first pair of walking legs with a large triangular projection near the middle of its distal edge (fig. 46D)— **Asellus communis** Say

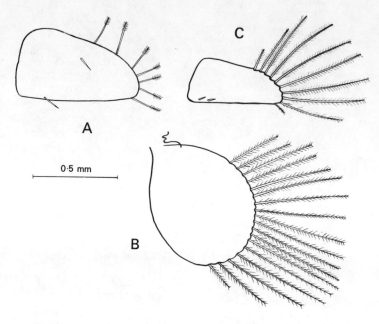

Fig. 48. Left "first" pleopods in ventral view of females of A, *Asellus cavaticus*; B, *A. aquaticus*; C, *A. meridianus* and *A. communis*.

FEMALES

5 (2) Pigment pattern on the dorsal surface of the head usually as in fig. 44B. "First" pair of pleopods rounded and overlapping each other (figs 45A, 48B)— **Asellus aquaticus** (L.)

— Pigment pattern on the dorsal surface of the head as in fig. 44B or C. "First" pair of pleopods trapezoidal or nearly so in shape and not overlapping each other (fig. 48C)— 6

6 Pigment pattern on the dorsal surface of the head usually as in fig. 44C (sometimes this pattern is masked by the underlying darker gut contents)— **Asellus meridianus** Racovitza

— Pigment pattern on the dorsal surface of the head usually as in fig. 44B— **Asellus communis** Say

Asellus aquaticus and *A. meridianus* are widely distributed throughout the British Isles, although they appear to be absent from some of the Western Isles and from the most northerly parts of Scotland. *A. meridianus* is the only species which has been recorded from the Isle of Man, Lundy, Skokholm, the Scilly Isles and many other offshore islands (Williams, 1962a, b). Both species may be found in habitats ranging from clear running water to stagnant polluted water. In general it appears that *A. aquaticus* is more tolerant of organically polluted waters than is *A. meridianus*. Williams (1963) discusses the ecological relationships between these two species, and Steel (1961) describes their life-cycles. *A. communis* is a species introduced from North America where it is widespread. It is recorded in Britain from a single locality, Bolam Lake, Northumberland (Williams, 1972; Sutcliffe, 1972a) but may well be present elsewhere. *A. cavaticus* is a subterranean animal which has been collected from a number of wells and caves in southern England and south Wales.

Henry & Magniez (1968) divided the genus *Asellus* into eight genera raising the subgenera *Proasellus* Dudich, containing the species *A. cavaticus* and *A. meridianus*, and *Conasellus* Stammer, containing *A. communis*, to generic level. The generic groupings proposed by Henry & Magniez (1968, 1970) have not been followed in this key as although the genus *Asellus* may need to be divided it is felt to be premature to do this until the American species can satisfactorily be included.

ACKNOWLEDGMENTS

The following are thanked for supplying information on the distribution of various species: Biological Records Centre; Cave Research Group of Great Britain; Mr D. G. Holland; Dr H. B. N. Hynes; Dr G. T. Jefferson; Mr T. Langford; Dr P. Maitland; Professor H. P. Moon; Dr B. W. Staddon. Mr A. E. Joyce and students kindly collected gammarids from northern Scotland and the Shetland Isles. Professor J. Stock (Amsterdam) and Dr R. J. Lincoln (British Museum, Natural History) are thanked for their advice on *Niphargus kochianus*. Dr Lincoln and Mr Holland made valuable comments on the key for *Gammarus*. Dr S. Husmann and Dr E. Serban kindly gave permission for some of their figures to be used. Mrs Jean Mackereth drew some of the new figures and Dr G. Fryer made some valuable comments on the first draft of the new edition. Mrs Rosalie Sutcliffe typed numerous drafts of the manuscript. Mr J. E. M. Horne edited the manuscript and supervised its conversion to the printed form.

REFERENCES

Abrahamsson, S. A. A. (1972). Fecundity and growth of some populations of *Astacus astacus* Linné in Sweden, with special regard to introductions in northern Sweden. *Rep. Inst. Freshwat. Res. Drottningholm,* **52**, 23-37.
Abrahamsson, S. [A. A.] (1973) Editor. *Freshwater crayfish.* 1st International Symposium, Austria, 1972. Lund, Studentlitteratur. pp. 252.
Allen, J. A. (1967). Crustacea: Euphausiacea and Decapoda, with an illustrated key to the British species. *Fauna of the Clyde Sea Area,* Millport. pp. 116.
Bartok, P. (1944). A *Bathynella chappuisi* fejlodes morphologiaja. *Acta Sci. Math. Nat. Koloszvar,* **21**, 1-46.
Bott, R. (1950). Die Flusskrebse Europas (Decapoda, Astacidae). *Abh. senckenb. naturforsch. Ges.* **483**, 1-36.
Bousfield, E. L. (1958). Freshwater amphipod crustaceans of glaciated North America. *Can. Fld Nat.* **72**, 55-113.
Bouvier, E.-L. (1940). Décapodes marcheurs. *Faune Fr.* **37**.
Calman, W. T. (1917). Notes on the morphology of *Bathynella* and some allied Crustacea. *Q. Jl microsc. Sci.* **62**, 489-514.
Chambers, M. R. (1973). Notes on the gammarid fauna of the Frisian Lake District following the invasion of the alien amphipod *Gammarus tigrinus* Sexton. *Bull. zool. Mus. Univ. Amsterdam* **3**, 1-6.
Crawford, G. I. (1935). *Corophium curvispinum* G. O. Sars, var. *devium* Wundsch, in England. *Nature, Lond.* **136**, 685.
Crawford, G. I. (1937a). A review of the amphipod genus *Corophium*, with notes on the British species. *J. mar. biol. Ass. U.K.* **21**, 589-630.
Crawford, G. I. (1937b). The fauna of certain estuaries in West England and South Wales, with special reference to the Tanaidacea, Isopoda and Amphipoda. *J. mar. biol. Ass. U.K.* **21**, 647-62.
Cukerzis, J. (1968). Interspecific relations between *Astacus astacus* L. and *A. leptodactylus* Esch. *Ekol. pol. (A),* **16**, 629-36.
Curry, A., Grayson, R. F. & Milligan, T. D. (1972). New British records of the semi-terrestrial amphipod *Orchestia cavimana.* *Freshwat. Biol.* **2**, 55-6+1 pl.
De Leersnyder, M. (1967). Influence de la salinité et de l'ablation des pedoncules oculaires sur la mue et sur le développement ovarien d'*Eriocheir sinensis* H. Milne Edwards. *Cah. Biol. mar.* **8**, 421-35.
Dennert, H. G., Dennert, A. L., Kant, P., Pinkster, S. & Stock, J. H. (1969). Upstream and downstream migrations in relation to the reproductive cycle and to environmental factors in the amphipod, *Gammarus zaddachi. Bijdr. Dierk.* **39**, 11-43.
Efford, I. E. (1959). Rediscovery of *Bathynella chappuisi* Delachaux in Britain. *Nature, Lond.* **184**, 558-9.
Fries, G. & Tesch, F. W. (1965). Der Einfluss der Massenvorkommens von *Gammarus tigrinus* Sexton auf Fische und niedere Tierwelt in der Weser. *Arch. FischWiss.* **16**, 133-50.
Fryer, G. (1950). A Yorkshire record of the amphipod *Orchestia bottae* (M.-Edw.). *Naturalist, Hull,* Oct.-Dec. 1950, 148.

Fryer, G. (1951). Notes on *Orchestia bottae* (M.-Edw.) and other non-marine Amphipoda. *Naturalist, Hull,* July-Sept. 1951, 105.

Furst, M. (1965). Experiments on the transplantation of *Mysis relicta* Lovén into Swedish lakes. *Rep. Inst. Freshwat. Res. Drottningholm* **46,** 79–89.

Gledhill, T. & Driver, D. B. (1964). *Bathynella natans* Vejdovsky (Crustacea; Syncarida) and its occurrence in Yorkshire. *Naturalist, Hull,* July-Sept. 1964, 104–6.

Gledhill, T. & Ladle, M. (1969). Observations on the life-history of the subterranean amphipod *Niphargus aquilex aquilex* Schiödte. *Crustaceana,* **16,** 51–6.

Glennie, E. A. (1967). The distribution of the hypogean Amphipoda in Britain. *Trans. Cave Res. Grp Gt Br.* **9,** 132–6.

Glennie, E. A. (1968). The discovery of *Niphargus aquilex aquilex* Schiödte in Radnorshire. *Trans. Cave Res. Grp Gt Br.* **10,** 139–40.

Goedmakers, A. (1972). *Gammarus fossarum* Koch, 1835: Redescription based on neotype material and notes on its local variation (Crustacea, Amphipoda). *Bijdr. Dierk.* **42,** 124–38.

Gordon, I. (1963). On the rostrum of the British crayfish, *Astacus pallipes* Lereboullet. *Crustaceana,* **5,** 234–8.

Gurney, R. (1923). Some notes on *Leander longirostris* M. Edwards, and other British prawns. *Proc. zool. Soc. Lond.* 1923, 97–123.

Hamond, R. (1971). The leptostracan, euphausiid, stomatopod, and decapod Crustacea of Norfolk. *Trans. Norfolk Norwich Nat. Soc.* **22,** 90–112.

Hazelton, M. (1974a). The fauna from some Irish caves. *Trans. Cave Res. Grp Gt Br.* **15,** 191–6.

Hazelton, M. (1974b). Irish vice county records of fauna collected from the hypogean and related zones. *Trans. Cave Res. Grp Gt Br.* **15,** 203–15.

Hazelton, M. (1974c). Hypogean fauna recorded from Ireland 1952–1971. *Trans. Cave Res. Grp Gt Br.* **15,** 225–52.

Henry, J.-P. & Magniez, G. (1968). Sur la systématique et la biogéographie des Asellides. *C.r. hebd. Séanc. Acad. Sci., Paris,* **267,** 87–9.

Henry, J.-P. & Magniez, G. (1970). Contribution à la systématique des Asellides (Crustacea Isopoda). *Annls Spéléol.* **25,** 335–67.

Hoestlandt, H. (1955). L'extension de l'*Eriocheir sinensis* H. M. Edw. (Crustacé décapode) en France depuis 1937. *C. r. Congr. Soc. sav Paris Sect. Sci.* **80,** 171–6.

Holland, D. G. (1976a). The distribution of the freshwater Malacostraca in the area of the Mersey and Weaver River Authority. *Freshwat. Biol.* (In press).

Holland, D. G. (1976b). The inland distribution of brackish-water *Gammarus* species in the area of the Mersey and Weaver River Authority. *Freshwat. Biol.* (In press).

Holmquist, C. (1959). Problems on marine-glacial relicts on account of investigations on the genus *Mysis*. Lund. pp. 270.

Holthuis, L. B. (1964). On the status of two allegedly European crayfishes, *Cambarus typhlobius* Joseph, 1880, and *Austropotamobius pallipes bispinosus* Karaman, 1961 (Decapoda, Astacidae). *Crustaceana,* **7,** 42–8.

Holthuis, L. B. (1967). Decapoda. *Limnofauna Europaea* (ed. J. Illies), Stuttgart. 189–92.

Husmann, S. (1968). Ökologie, Systematik und Verbreitung zweier in Norddeutschland sympatrisch lebender *Bathynella*-Arten (Crustacea, Syncarida). *Int. J. Speleol.* **3,** 111–45+pl. 24–36.

Hynes, H. B. N. (1951). Distribution of British freshwater Amphipoda. *Nature, Lond.* **167**, 152.

Hynes, H. B. N. (1954a). The ecology of *Gammarus duebeni* Lilljeborg and its occurrence in fresh water in western Britain. *J. Anim. Ecol.* **23**, 38–84.

Hynes, H. B. N. (1954b). The identity of *Gammarus tigrinus* Sexton 1939. *Nature, Lond.* **174**, 563.

Hynes, H. B. N. (1955a). Distribution of some freshwater Amphipoda in Britain. *Verh. int. Verein. theor. angew. Limnol.* **12**, 620–8.

Hynes, H. B. N. (1955b). The reproductive cycle of some British freshwater Gammaridae. *J. Anim. Ecol.* **24**, 352–87.

Karaman, M. S. (1962). Ein Beitrage zur Systematik der Astacidae (Decapoda). *Crustaceana*, **3**, 173–91.

Kinne, O. (1954). Die *Gammarus*-Arten der Kieler Bucht. *Zool. Jb. (Syst.)* **82**, 405–96.

Laurent, P. J. (1960). Systématique des Astacidae de France. *Annls Stn cent. Hydrobiol. appl.* **8**, 265-80.

Laurent, P. J. & Suscillon, M. (1962). Les Écrevisses en France. *Annls Stn cent. Hydrobiol. appl.* **9**, 333–97.

Lund, H. M. (1969). Krepsen i Norge, dens miljokrav og okonomiske verdi. *Fauna, Oslo*, **22**, 177–88.

Maitland, P. S. (1962). *Bathynella natans*, new to Scotland. *Glasg. Nat.* **18**, 175–6.

Moon, H. P. (1953). A re-examination of certain records for the genus *Asellus* (Isopoda) in the British Isles. *Proc. zool. Soc. Lond.* **123**, 411–17.

Moon, H. P. (1970). *Corophium curvispinum* (Amphipoda) recorded again in the British Isles. *Nature, Lond.* **226**, 976.

Moriarty, C. (1973). A study of *Austropotamobius pallipes* in Ireland. In *Freshwater Crayfish* (ed. S. Abrahamsson), 56–67. Lund.

Naylor, E. (1972). British marine isopods. *Synopses Br. Fauna*, (N.S.) **3**, pp. 86.

Nijssen, H. & Stock, J. H. (1966). The amphipod, *Gammarus tigrinus* Sexton, 1939, introduced in the Netherlands (Crustacea). *Beaufortia*, **13**, 197–206.

Noodt, W. (1965). Natürliches System und Biogeographie der Syncarida (Crustacea Malacostraca). *Gewäss. Abwäss.* **37/38**, 77–186.

Ökland, K. A. (1969). On the distribution and ecology of *Gammarus lacustris* G.O. Sars in Norway, with notes on its morphology and biology. *Nytt Mag. Zool.* **17**, 111–52.

Pinkster, S. (1970). Redescription of *Gammarus pulex* (Linnaeus, 1758) based on neotype material (Amphipoda). *Crustaceana*, **18**, 177–86.

Pinkster, S. (1972). On members of the *Gammarus pulex*-group (Crustacea–Amphipoda) from western Europe. *Bijdr. Dierk.* **42**, 164–91.

Pinkster, S. (1973). The *Echinogammarus berilloni*-group, a number of predominantly Iberian amphipod species (Crustacea). *Bijdr. Dierk.* **43**, 1–38.

Pinkster, S., Dennert, A. L., Stock, B. & Stock, J. H. (1970). The problem of European freshwater populations of *Gammarus duebeni* Lilljeborg, 1852. *Bijdr. Dierk.* **40**, 116–47.

Racovitza, E. G. (1919). Notes sur les Isopodes. 1—*Asellus aquaticus* auct. est une erreur taxonomique. 2—*Asellus aquaticus* L. et *A. meridianus* n.sp. *Archs Zool. exp. gén.* **58**, 31–43.

Reid, D. M. (1939). On the occurrence of *Gammarus duebeni* (Lillj.) (Crustacea, Amphipoda) in Ireland. *Proc. R. Ir. Acad. (B)*, **45**, 207–14.

REFERENCES

Reid, D. M. (1947). Talitridae (Crustacea, Amphipoda). *Synopses Br. Fauna,* No. 7, pp. 25.

Roux, A. L. (1967). Les Gammares du groupe *pulex* (Crustacés–Amphipodes). Essai de systématique biologique. *Thès. Fac. Sci. Dr. Univ. Lyon,* **447**, 1–172.

Rygg, B. (1974). Identification of juvenile Baltic gammarids (Crustacea, Amphipoda). *Annls zool. fenn.* **11**, 216–19.

Schellenberg, A. (1942). Krebstiere oder Crustacea 4: Flohkrebse oder Amphipoda. *Tierwelt Dtl.* **40**, 1–252.

Scourfield, D. J. (1940). Note on the difference in the coloration of the head in *Asellus aquaticus* and *Asellus meridianus*. *Essex Nat.* **26**, 268–70.

Segerstråle, S. G. (1947). New observations on the distribution and morphology of the amphipod *Gammarus zaddachi* Sexton, with notes on related species. *J. mar. biol. Ass. U.K.* **28**, 219–44.

Segerstråle, S. G. (1954). The freshwater amphipods *Gammarus pulex* (L.) and *Gammarus lacustris* G. O. Sars, in Denmark and Fennoscandia—a contribution to the late and postglacial immigration history of the aquatic fauna of northern Europe. *Commentat. biol.* **15**, 1, 1–91.

Segerstråle, S. G. (1957). On the immigration of the glacial relicts of northern Europe, with remarks on their prehistory. *Commentat. biol.* **16**, 16, 1–117.

Segerstråle, S. G. (1966). Adaptational problems involved in the history of the glacial relics of Eurasia and North America. *Revue roum. Biol. (Zool.),* **11**, 59–66.

Serban, E. (1966a). Contribution à l'étude de *Bathynella* d'Europe; *Bathynella natans* Vejdovsky, un dilemme à résoudre. *Int. J. Speleol.* **2**, 115–32+pls. 29–35.

Serban, E. (1966b). Nouvelles contributions à l'étude de *Bathynella (Bathynella) natans* Vejd. et *Bathynella (Antrobathynella) stammeri* (Jakobi). *Int. J. Speleol.* **2**, 207–21+pl. 43–8.

Serban, E. (1970). A propos du genre *Bathynella* Vejdovsky (Crustacea Syncarida). In *Livre du centenaire Emile G. Racovitza 1868–1968,* (ed. T. Orghidan), 265-73. Bucarest.

Serban, E. (1972). *Bathynella* (Podophallocarida Bathynellacea). *Trav. Inst. Speol. Emile Racovitza,* **11**, 11–225.

Serban, E. (1973). Sur les Bathynellidae (Podophallocarida Bathynellacea) de l'Italie: *Bathynella ruffoi* nov. sp. et *Bathynella lombardica* nov. sp. *Memorie Mus. civ. Stor. nat. Verona,* **20**, 17–37.

Serban, E. & Gledhill, T. (1965). Concerning the presence of *Bathynella natans stammeri* Jakobi (Crustacea: Syncarida) in England and Rumania. *Ann. Mag. nat. hist.* (Ser. 13), **8**, 513–22.

Sexton, E. W. (1913). Description of a new species of brackish-water *Gammarus* (*G. chevreuxi*, n.sp.). *J. mar. biol. Ass. U.K.* **9**, 542–55.

Sexton, E. W. (1924). The moulting and growth-stages of *Gammarus*, with description of the normals and intersexes of *G. chevreuxi*. *J. mar. biol. Ass. U.K.* **13**, 340–401.

Sexton, E. W. (1939). On a new species of *Gammarus (G. tigrinus)* from Droitwich District. *J. mar. biol. Ass. U.K.* **23**, 543–52.

Sexton, E. W. (1942). The relation of *Gammarus zaddachi* Sexton to some other species of *Gammarus* occurring in fresh, estuarine and marine waters. *J. mar. biol. Ass. U.K.* **25**, 575–606.

Sexton, E. W. & Clark, A. R. (1937). A summary of the work on the amphipod *Gammarus chevreuxi* Sexton carried out at the Plymouth laboratory (1912–1936). *J. mar. biol. Ass. U.K.* **21,** 357–414.

Sexton, E. W. & Spooner, G. M. (1940). An account of *Marinogammarus* (Schellenberg) gen. nov. (Amphipoda) with a description of a new species, *M. pirloti. J. mar. biol. Ass. U.K.* **24,** 633–82.

Shoemaker, C. R. (1942). Notes on some American freshwater amphipod crustaceans and descriptions of a new genus and two new species. *Smithson. misc. Collns,* **101,** No. 9.

Spooner, G. M. (1947). The distribution of *Gammarus* species in estuaries. Part I. *J. mar. biol. Ass. U.K.* **27,** 1–52.

Spooner, G. M. (1951). On *Gammarus zaddachi oceanicus* Segerstråle. *J. mar. biol. Ass. U.K.* **30,** 129–47.

Spooner, G. M. (1957). In "Plymouth Marine Fauna", 3rd Edition. Marine Biological Association, Plymouth.

Spooner, G. M. (1961). *Bathynella* and other interstitial Crustacea in southern England. *Nature, Lond.* **190,** 104–5.

Steel, E. A. (1961). Some observations on the life history of *Asellus aquaticus* (L.) and *Asellus meridianus* Racovitza (Crustacea: Isopoda). *Proc. zool. Soc. Lond.* **137,** 71–87.

Stock, J. H. (1967). A revision of the European species of the *Gammarus locusta*-group (Crustacea, Amphipoda). *Zool. Verh., Leiden,* **90,** 1–56.

Stock, J. H. (1969). *Rivulogammarus,* an amphipod name that must be rejected. *Crustaceana,* **17,** 106–7.

Stock, J. H. & Gledhill, T. (in press). The *Niphargus kochianus*-group in north-western Europe. 3rd int. Colloquium on *Gammarus* and *Niphargus,* Schlitz, Sept. 1975. *Crustaceana.*

Stock, J. H. & Pinkster, S. (1970). Irish and French freshwater populations of *Gammarus duebeni* subspecifically different from brackish water populations. *Nature, Lond.* **228,** 874–5.

Straskraba, M. (1972). L'etat actuel de nos connaissances sur le genre *Niphargus* en Tchecoslovaquie et dans les pays voisins. *Memorie Mus. civ. Stor. nat. Verona fuori serie* N. **5,** 35–46.

Stringer, G. E. (1967). Introduction of *Mysis relicta* Lovén into Kalamalka and Pinaus Lakes, British Columbia. *J. Fish. Res. Bd Can.* **24,** 463–5.

Sutcliffe, D. W. (1967). A re-examination of observations on the distribution of *Gammarus duebeni* Lilljeborg in relation to the salt content in fresh water. *J. Anim. Ecol.* **36,** 579–97.

Sutcliffe, D. W. (1968). Sodium regulation and adaptation to fresh water in gammarid crustaceans. *J. exp. Biol.* **48,** 359–80.

Sutcliffe, D. W. (1970). Experimental populations of *Gammarus duebeni* in fresh water with a low sodium content. *Nature, Lond.* **228,** 875–6.

Sutcliffe, D. W. (1971a). Sodium influx and loss in freshwater and brackish-water populations of the amphipod *Gammarus duebeni* Lilljeborg. *J. exp. Biol.* **54,** 255–68.

Sutcliffe, D. W. (1971b). Regulation of water and some ions in gammarids (Amphipoda). I. *Gammarus duebeni* Lilljeborg from brackish water and fresh water. *J. exp. Biol.* **55,** 325–44.

REFERENCES

Sutcliffe, D. W. (1972a). Notes on the chemistry and fauna of water-bodies in Northumberland with special emphasis on the distribution of *Gammarus pulex* (L.), *G. lacustris* Sars and *Asellus communis* Say (new to Britain). *Trans. nat. Hist. Soc. Northumb.* **17**, 222–48.

Sutcliffe, D. W. (1972b). An examination of subspecific differences in the merus of the fifth walking leg of the amphipod *Gammarus duebeni* Lilljeborg. *Freshwat. Biol.* **2**, 203–16.

Sutcliffe, D. W. (1974). On *Gammarus* from fresh waters in the islands of Orkney and Shetland. *Crustaceana*, **27**, 109–11.

Sutcliffe, D. W. & Carrick, T. R. (1973a). Studies on mountain streams in the English Lake District. I. pH, calcium and the distribution of invertebrates in the River Duddon. *Freshwat. Biol.* **3**, 437–62.

Sutcliffe, D. W. & Carrick, T. R. (1973b). Studies on mountain streams in the English Lake District. III. Aspects of water chemistry in Brownrigg Well, Whelpside Ghyll. *Freshwat. Biol.* **3**, 561–8.

Sutcliffe, D. W. & Shaw, J. (1968). Sodium regulation in the amphipod *Gammarus duebeni* Lilljeborg from freshwater localities in Ireland. *J. exp. Biol.* **48**, 339–58.

Svärdson, G. (1965). The American crayfish *Pacifastacus leniusculus* (Dana) introduced into Sweden. *Rep. Inst. Freshwat. Res. Drottningholm.* **46**, 90–4.

Svärdson, G. (1972). The predatory impact of eel (*Anguilla anguilla* L.) on populations of crayfish (*Astacus astacus* L.). *Rep. Inst. Freshwat. Res. Drottningholm.* **52**, 149–91.

Tattersall, W. M. (1930). *Asellus cavaticus* Schiödte, a blind isopod new to the British fauna, from a well in Hampshire. *J. Linn. Soc. (Zool.)* **37**, 79–91.

Tattersall, W. M. & Tattersall, O. S. (1951). *The British Mysidacea.* London (Ray Society). pp. 460.

Thomas, W. (1974). A note on the rostrum of the English crayfish. *Trans. Kent Fld Club* **5**, 88–9.

Thomas, W. & Ingle, R. (1971). The nomenclature, bionomics and distribution of the crayfish, *Austropotamobius pallipes* (Lereboullet) (Crustacea, Astacidae) in the British Isles. *Essex Nat.* **32**, 349–60.

Unestam, T. (1969). Resistance to the crayfish plague in some American, Japanese and European crayfishes. *Rep. Inst. Freshwat. Res. Drottningholm*, **49**, 202–9.

Vejdovsky, F. (1907). On some freshwater amphipods: the reduction of the eye in a new gammarid from Ireland. *Ann. Mag. nat. Hist.* **20**, 227–45.

Williams, W. D. (1962a). The genus *Asellus* in Britain. *Nature, Lond.* **193**, 900–1.

Williams, W. D. (1962b). The geographical distribution of the isopods *Asellus aquaticus* (L.) and *A. meridianus* Rac. *Proc. zool. Soc. Lond.* **139**, 75–96.

Williams, W. D. (1963). The ecological relationships of isopod crustaceans *Asellus aquaticus* (L.) and *A. meridianus* Rac. *Proc. zool. Soc. Lond.* **140**, 661–79.

Williams, W. D. (1972). Occurrence in Britain of *Asellus communis* Say, a North American freshwater isopod. *Crustaceana Suppl.* **3**, 134–8.

INDEX

Names in parentheses are synonyms. Page numbers in italics are key references. Page numbers in bold type indicate illustrations.

(*Antrobathynella*) 7
Asellus 6, **58**, *59-65*
    aquaticus 59, **60**, *61* ♂, **62**, 63, 64 ♀, 65
    cavaticus 60, **62**, 63, 64, 65
    communis 60, *62* ♂, 63, 64, 65 ♀
    meridianus 59, **60**, *62* ♂, 63, 64, 65 ♀
Astacus astacus 14, 16, *20*, **21**
    colchicus 16
    (*fluviatilis*) 20
    leptodactylus 14, 16, *20*, **21**
    (*nobilis*) 20
    pachypus 16
    (*pallipes*) 14, 18
    (*torrentium*) 16
Austropotamobius pallipes 14, **15**, 16, *18* (+ subspp.), **19**
    torrentium 14, 16, **17**
Bathyonyx de Vismesi 30
Bathynella 6, *7-11*
    natans 7, *8*, **9**, **10**, **11**
    stammeri 7, *8*, **9**, **10**, **11**
(*Cambaroides schrenckii*) 18
(*Cambarus affinis*) 16
(*Conasellus*) 65
Corophium **27**
    curvispinum 26
    spongicolum 26
Crangon crangon 12
(*Crangonyx gracilis*) 28
Crangonyx pseudogracilis 28, **29**
    subterraneus 28, **29**
Echinogammarus berilloni 30, **31**
Eriocheir sinensis 14, **15**
(*Eucrangonyx gracilis*) 28
    (*vejdovskyi*) 28
Gammarus 12, 24, **25**, 30, *32-51*, 33
    chevreuxi 32, 33, *48*, **49**
    crinicornis 50
    duebeni 32, *35*, 37, *38*, 39, *40* (+ subspp.), **41**, **45**, 46, 50

fasciatus 48
fossarum 38
inaequicauda 50
insensibilis 50
lacustris 32, **35**, *36*, 37, *38*, 39
locusta 32, 43, 46, *50*, **51**
oceanicus 32, 33, 43, 46, *50*, **51**
pulex 32, 35 *36*, **36**, 37, *38* (+ subspp.), **39**
salinus 32. 33, 43, 46, 47, 50
tigrinus 32, 33, *37*, 39, **41**, 45, *48*, **49**
wautieri 38
zaddachi 32, 33, 37, 38, 39, **41**, 43, 44, **45**, 46, 47, 50
Jaera 58
    nordmanni **58**, 59
(*Leander*) 12
Marinogammarus 12, *30*
    finmarchicus 30
Mysis relicta **23**, *23*
Neomysis integer 23
    (*vulgaris*) 23
Niphargellus 28, *52*
    glenniei 52, **53**, **55**
Niphargus 28, *52*
    aquilex **52**, 55, *56* (+ subspp.), **57**
    fontanus **55**, 56, **57**
    kochianus **53**, *54*, **55**
    kochianus irlandicus 56
    kochianus kochianus **53**, *54*, 55
(*Orchestia bottae*) 26
Orchestia cavimana 26, **27**
Orconectes limosus 14, 16, **17**
Pacifastacus leniusculus 14
Palaemon longirostris 12
Palaemonetes varians 12, **13**
(*Potamobius pallipes*) 14
(*Proasellus*) 65
(*Rivulogammarus*) 38
Sphaeroma 58, 59
    hookeri **58**, 59
    rugicauda 59